D1356459

DER SPIEGEL
VII. JAHRGANG
NR. 50
VERLAGSORT HAMBURG
HÄSSLICHKEIT VERKAUFT SICH SCHLECHT
Kreuzzug des guten Geschmacks: Formgestalter Loewy (siehe „Industrie")

Raymond Loewy

PAUL JODARD

TREFOIL PUBLICATIONS, LONDON

Published by Trefoil Publications Ltd
7 Royal Parade, Dawes Rd
London SW 6

First published 1992

ISBN 0 86294 150 4

Design by the Armelle Press after an original design by Elizabeth van Amerongen
Typeset by Jamesway Graphics
Manufactured in Singapore

contents

For A., for M. and for A.

heroes of the design century

More that half a century ago, on January 18th 1933, *Punch* published a menacing cartoon. It shows a figure described as an employer seated at a desk. Standing in front of him is a monstrous mechanical man. In the caption the robot is saying; 'Master, I can do the work of fifty men.' The employer replies; 'Yes, I know that. But *who is to support the fifty men*?'

The ambivalent answer to the cartoon employer's question has been the story of the 20th century. Ever since the Industrial Revolution liberated the developed nations from the agricultural economy of scarcity, the prospect of automation has embodied a threat as well as a promise. A threat of mechanised poverty, oppression, starvation and eventually genocide to set against the promise of automated wealth, freedom, plenty and procreation.

But as we look back over the 20th century, it is clear that one activity above all others has come to dominate the shape of the machine-made world. Reality in the late 20th century is *designed*, not shaped by use or custom. In this sense it is already correct to call the 20th century the century of design, for design has been its response to the agonized question of the employer in the *Punch* cartoon.

From diffuse and uncertain beginnings design has become the only human activity that can still promise to connect the support of the fifty men robbed of their work by machines to the productive capabilities of the robot. By perpetually extending the mechanical man in the direction of the human economy, the designers of the 20th century created a world that had never before existed in history. They found ways to close the interface between human consciousness and the man-made material world to a magical fit. Where once technology killed, now it fits. Where once the machine was feared, now it is embraced – and both transformations were carried out by design.

That the designers of the 20th century really have connected the necessity of support with the abundance of invention is a proposition that requires little proof. Even the most extravagant interpretation of human needs in 1933 could scarcely have included one tenth of the goods and services whose universal distribution has today been rendered normal by the activity of design. The multifarious inanimate energy slaves of the late 20th century city would have struck the employer of 1933 as a nightmare of implied social support. Fifty robots instead of one! Who could support the 2,500 men thus thrown out of work? Nor would the answer to this question have seemed credible: that tasks would be created that sixty years ago did not exist. These new tasks, and the purchasing power they have created, are the engines that power the expanding universe of design and production today. The symbiotic phenomenon of design and economic life now exists at a level far beyond the wildest dreams of the industrialists of sixty years ago. Design now sustains economic activity. Already, sixty years after that *Punch* cartoon, design has become the key to the new material world that can support fifty times fifty men. Design is the interdenominational networking of form that makes sense of consumption. From the ten-speed racer to the VCR to the cellular telephone to the workstation to the communications satellite, design is the adhesive that binds all to the global economy. In the end it is design that makes the buying and selling of information as intelligible as it once was to buy and sell slaves and cotton.

But if the new role of design as the engine of consumption is as evolutionarily important as this, how are we to understand its individual workings? At the level of production economics, designers seem no more than a human sub-species, like bees, who work out a set of general principles without character or individuality. Indeed today, in investment terms, it has become possible to see the design profession in this way, but this is a very recent development.

'Design Heroes' is a series about the individuals who shaped this now homogeneous world of post-industrial design. Men and women who did not resemble the stereotype of today's designers, but who nonetheless created their world of present possibilities for them. Although they seldom worked entirely alone, these 'Design Heroes' established themselves as individual talents, rather than as members of packs like today's design consultancies; packs that are identified by strings of names, cryptic initials or acronyms. These interchangeable box-suited Ray-Ban figures driving BMWs may be quoted on the USM or listed on Stock Exchanges all over the world, but they are not 'Design Heroes', yet. The classification is reserved for those bold individuals who staked out the first claims in the virgin territory that the corporate designer of today is content to methodically comb and recomb, with the certainty and conviction of the 1933 robot.

Who were these prime individuals, the creators of a new genre of creative human being? The question is more difficult to answer than it might appear, not least because it is only in the last few years that the synergetic importance of design and economics has been popularly grasped.

The the 20th century design always existed, but for years it was not recognized as an unitary phenomenon and its practitioners did their work under different titles with different degrees of status. Sometimes what we would call designers today were 'chief draughtsmen', sometimes 'engineers', sometimes 'inventors', sometimes 'craftsmen', sometimes 'artists', sometimes 'amateur constructors'. In fact the best of them were always Heroes if, consciously or unconsciously, they worked to the dictum of Nietzsche: 'Need is not the reason for something to come into existence, it is an effect of what has already come to be'.

This series of short biographies is an attempt to tell, through the lives of a number of great 20th century designers, how the question posed by the *Punch* cartoon was, and is, being answered. The 'Design Heroes' are men and women who somehow and in some way overdrew on the bank of invention, and in doing so revealed something of the inner mechanism of the creative individual under stress and thus helped to define the elusive modern term 'designer'. All these individuals have been chosen because, in widely different ways, their lives and their works deliver the essence of design as a vital human activity.

We know that it was determination, stamina, endurance beyond the call of reason that created the 'Heroes' of exploration; the 'Heroes' of warfare; the 'Heroes' of speed and flight. design too makes its calls upon determination, stamina and endurance. 'Design Heroes', like all heroes, are individuals who have been beyond the point of reasonable withdrawal. They have suffered for their work and their convictions. They have overstepped the bounds of conventional behaviour in order not to relinquish the creative integrity of their work.

We know from history that it was not science, but design that created the first engines to pump water; the first mechanical tools to lift rock, bore tunnels and bridge rivers; the first ships that could sail against the wind. Design too created the man-made environment and defined the limits of the dreams of what might still be possible within it. 'Design Heroes' is not a series about the great inventors of 19th century technology. It is about the generation that grew up with the elements of the modern world, the car, the passenger aeroplane, the spacecraft and the computer. Men like Richard Buckminster Fuller, whose long life encompassed the history of flight and the history of prefabrication. Raymond Loewy, who designed railway locomotives as well as the interiors of NASA spacecraft. Harley Earl, the creator of the surrealistic finned monsters of the post-World War Two American automobile industry. Ettore Sottsass, who worked on the first Olivetti computers before he broke free from the constraints of Modernism altogether and entered a revolutionary new creative world of furniture design. Colin Chapman, who founded a high-performance automobile legend that he used every resource, even forbidden ones, to keep out of the hands of corporate predators until he died. Tom Karen, who turned a three-wheeled van, a defunct sports car prototype, and the design for the first ever car to be produced in Turkey, into a sought-after design that was driven by Royalty.

Through the lives and works of designers like these, the series 'Design Heroes' will probe into our understanding of what those men did who truly learned how to make more production out of less work than had ever been done before – by design.

Martin Pawley
October 1991

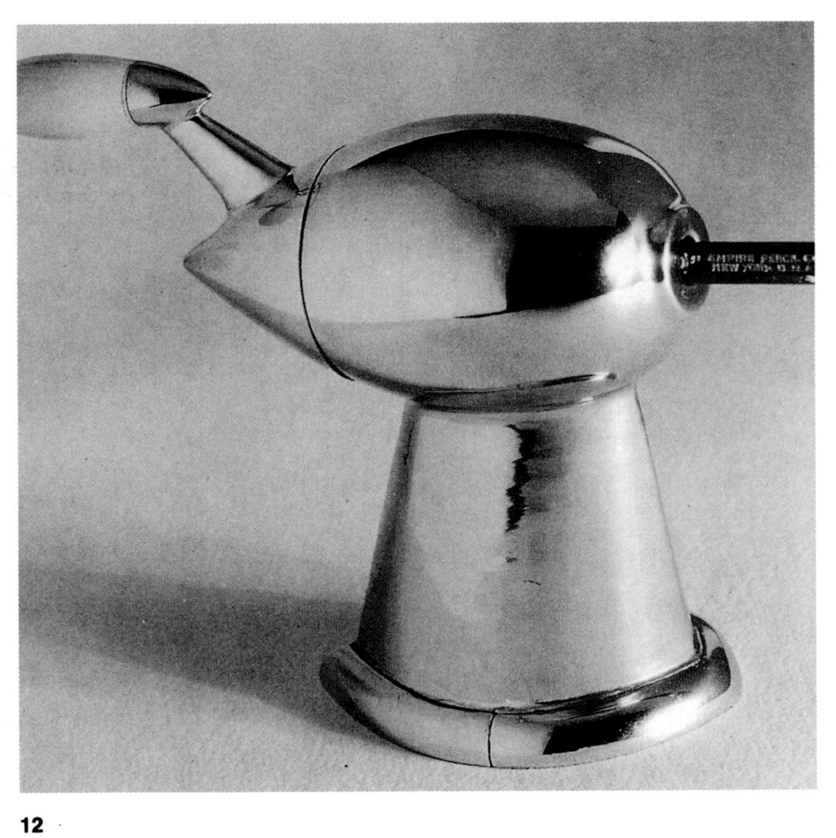

the paradoxical pioneer

'There was something about the American landscape that made you feel that it had just been discovered... *America*: it suggests to us even now a landscape unfamiliar and wild... There were adventures and new forms of life over there.'

Edmund Wilson, *Memoirs of Hecate County*

Raymond Loewy spent an afternoon in 1962 with coloured paper and crayons, on the floor of the Oval Office in the White House, helping President John F. Kennedy choose a new colour scheme for Air Force One, the President's official aircraft. The drawing for the final colour scheme is now in the Library of Congress: the purchase of these and other Loewy drawings and their donation to the Library in 1987 inaugurated the Library's Design Collection. Loewy – who had died the previous year at the age of 93 – would have liked the fact of being first. As a longtime colleague Evert Endt puts it in the recent *Raymond Loewy* exhibition catalogue, he had a 'genial knack of pushing himself to the front'.

The streamlined pencil sharpener: a design by Loewy that was never manufactured. Air resistance is not a normal problem in sharpening pencils, but the result still looks lovely.

The drawing decks the aircraft out in patriotic red, white and blue, set off with a green and silver background. It is at best a competent piece of work, with little or no drama and not much originality, and would pass unnoticed

but for the 'hands-on' involvement of the famous client. The client's choice of designer is also interesting: Loewy had come to prominence in the years before the Second World War, and a younger generation of designers such as Charles Eames and Eliot Noyes would be more in tune with the Kennedy era. In his own account of the meeting Loewy does not make it clear how the invitation came about, but clarity was not always a strong point in Loewy's memoirs. He suggests, for example, that the meeting with President Kennedy led directly to an invitation to work with NASA on the Skylab project, though in fact the invitation did not come for another five years and did not come directly from NASA, even then. But the error might be due to enthusiasm, something which Loewy never seemed to have in short supply. His two biographical books *Never Leave Well Enough Alone* (1951) and *Industrial Design* (1979) are brimming with jaunty good humour and effervescent with famous names. Everyone who is good is great and Loewy and his boys have the answer to their problems. The first book, which was published in New York in 1951, contains a racy account

Trains and planes: Air Force One, in the livery designed by Loewy, assisted by President John F. Kennedy *(facing)* and an early (1911) sketch of a steam train.

... and automobiles: title page decoration for *Industrial Design (below)*.

of Loewy's childhood, war experience and first years in the USA, interspersed with *intermezzo* explanations of basic design ideas and behind-the-scenes looks at life in the design office. The subtitle reads 'The personal record of an industrial designer from lipsticks to locomotives', and the unusual format and original typography ensured the book a certain success: it

The young Raymond Loewy, before and during *(facing)* the First World War.

was translated into several languages, including Japanese and Arabic. *Industrial Design* takes advantage of modern printing technology to integrate illustrations and text, and so reads more as a series of magazine spreads on different topics, with Loewy's breathless prose running through both books. As the endless photographs of Raymond with this or that product and this or that celebrity flash by, the reader has to fight the impression that every paragraph begins with the word 'I'. Everything moves forward with the relentless dash of a steam locomotive, with Loewy on the footplate offering dapper asides on the American dream, on the role of the designer, on the future of transport, shopping, living: 'how far ahead can the designer go, stylewise?'

Indeed Loewy's other book – more of a photo-essay – is *The Locomotive*, first published in 1937: it is dedicated to Engine 3768 of the Pennsylvania Railroad with its designer's 'heartiest wishes for a fast and brilliant career'. Several of the illustrations show the designer in front of the streamlined steam locomotive, and with the diesel-electric engine he designed for the same Railroad. The captions to the illustra-

The cover of Loewy's 1937 book *Locomotive*.

tions comment on the aesthetics of other locomotive designs, talking of the 'glamour of the Steam Age' and 'railroad romance'. The book says nothing about the development of locomotive engineering – whether in steam or diesel electric traction – and the comments are almost always about the appearance of externals.

This highlights one of the common criticisms of Loewy's work as a designer, that he was a stylist, more concerned with getting the outside of a product to look right for the marketplace – or for the client – than with serious issues of design quality. His talent for self-promotion has done nothing to deflate such criticisms. And yet his importance in the evolution of professional industrial design in America is undoubted, and it was his very enthusiasm and flair for getting himself and his designs into the limelight that helped to bring the value of design to the notice of American industry, and establish the profession of industrial designer as an essential arm of manufacturing business. Understanding this paradox is much of the key to understanding Raymond Loewy.

Raymond Loewy, the youngest of three brothers, was born in Paris in 1893, into a reasonably prosperous middle-class family: his Viennese-born father was the editor of a business journal, and his mother was French. His own description of his schooldays suggests he was more interested in beauty than science: he describes at great length the pleasures of travelling by steam train to school each day and the sights and people he encountered, and he was endlessly sketching and making drawings – especially of trains and cars. A photograph of Loewy taken in 1907 shows an elegant young man – seeming older than his 14 years – with a fashionable *noeud papillon* and centre parting, looking firmly at the camera. He served in the First World War from the beginning – he was doing his military service in the French Army when war was declared – and finished as a captain in the artillery. His description of bringing back from leave to the trenches 'wall paper and coloured silks rather than sausages and brandy' suggest that Loewy maintained his aesthetic sense under fire: he redesigned his own uniform, as well. But he worked hard as an artillery officer, and was three times decorated for gallantry, once for going out repeatedly to repair telephone lines under fire.

Both his elder brothers survived the war, but the Spanish flu epidemic of 1918 was to kill his parents. Max and Georges, his brothers, emigrated to America – in the case of Georges his experience as a doctor in the War brought him an invitation to work at the Rockefeller Institute in New York with poison gas victims. With nothing to inherit from his parents, and no remaining ties in Europe, Raymond, in his customized uniform and with forty dollars from his war service gratuity, set off to New York too, aboard the liner France, in September, 1919. On board ship, by his own account, a drawing he made of a girl on the sun deck was auctioned for the benefit of a fund for shipwrecked sailors, and the purchaser, Sir Henry Armstrong, the British consul in New York, offered to put him in touch with Condé Nast and other magazine editors in New York: until then, Loewy had had 'no thoughts of such a career'. These introductions were a turning-point.

His first impressions of the bustle, clangour and sheer size of New York have often been quoted : 'the first impact was brutal. The giant scale of all things. Their ruggedness, their bulk, were frightening. Subways were thundering masses of sinister force, streetcars were monstrous and clattering hunks of rushing cast iron...' In a short time, however, he had adapted to life in the city, meeting people and making friends, selling his drawings, and moving in a group of fashion and press contacts who gave him work as an illustrator for advertising and fashion magazines, where he was known as 'the French artist', to his own evident satisfaction, for his reputation was boosted by the interest in French applied arts after the 1925 Paris Art Deco exhibition: many of the works were subsequently exhibited in New York and the press described Loewy as among the foremost Art Deco artists. His work on fashion illustrations for Harper's Bazaar and other magazines led to window dressing for

The French artist at work: the advertisement for Renault appeared in 1928.

Bonwit Teller and Macy's – though by his own account his first attempt to bring chic to Macy's previously cluttered windows met with such disapproval from management that he resigned rather than wait to be sacked.

His first industrial design commission came from the contacts he had made in the fashion and magazine world. The British industrialist, Sigmund Gestetner, whose company manufactured duplicating machines, had seen some of his work while in New York, and was introduced to Loewy. Gestetner wanted to improve the sales of his machine which, according to Loewy, looked ugly and smelt bad. With

characteristic dash, Loewy offered to reshape the machine within three days, and only to charge his costs of 500 dollars rather than a 2,000 dollar fee if the new design was not acceptable. Gestetner agreed to this sporting proposition and sent round the machine: Loewy sent his man out for a tarpaulin and 100 dollars worth of modelling clay, and set to work on his living room floor (protected by the tarpaulin) to encase the one in the other.

The original machine looked like a refugee from the Great Exhibition, its exposed mechanism arbitrarily decorated with curved bars and perched unsteadily on a boxy cabinet with splay feet, the whole stained with a sticky grey dust of paper and ink and 'covered with a mysterious bluish down that looked like the mold on tired Gorgonzola'. Loewy's solution was to straighten the feet, round out and raise the cabinet, and enclose the machinery in a smooth even casing, broken only by essential controls, which kept the dust – and the smell – well inside. The design was acceptable, and Loewy got his fee, together with a retainer to act as

Other advertising clients included Nieman Marcus *(facing)* and Bonwit Teller *(right)*

Gestetner's designer (an account that the London office of Loewy International still held through the 1980s). More importantly, he got a taste for industrial design, and realized that here was a possible career. In his own words (quoted in the recent *Loewy* catalogue) 'the designer's life is an agreeable one: I do what I like doing.'

Yet in this first project the contradictions and differences of viewpoint that beset an assessment of Loewy's achievements are already

Loewy's first industrial design commission: the Gestetner duplicating machine before *(left)* and after redesign.

apparent. The work that Loewy did involved no fundamental rethinking of the way the machine worked, no application of principles about appropriate form, and little saving in production costs or energy use (the main saving was in using plain screws instead of brass ones, as they were no longer visible under the new cover). It was an unashamed face-lift, made for the purposes of the marketplace, after what

seems almost a bet with the client. On the other hand, as a redesign, it represented a trend, that was to move such products away from industrial machinery and towards office furniture, a reflection of the growing importance of office work, in the country that was busy inventing the skyscraper. The differences between the empirical American approach to design and the theoretical European one were here put into

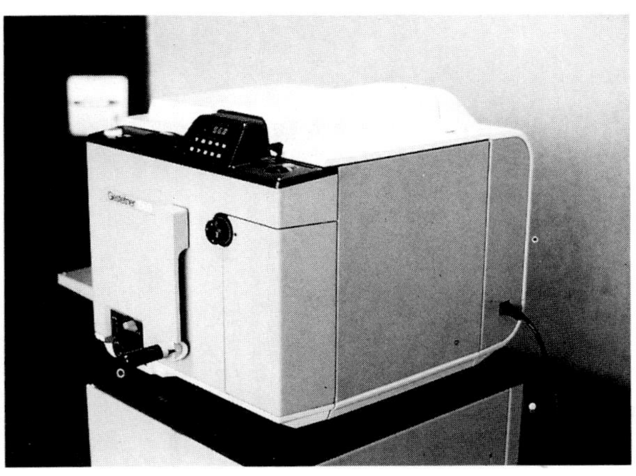

Later work for Gestetner included the Robin labelling system *(above)* and the 1566 duplicator *(below)* both designed in the London office.

clear contrast, and the word styling began its descent into the pejorative. European design looked back to the work of Peter Behrens for AEG, to the principles of the Werkstatte, and beyond to the ideas of William Morris and the Arts and Crafts Movement: design was almost seen as a moral activity, moving society towards better ends through honesty and the application of principles. The relationships between beauty and utility, and between work and craft, had little resonance in the younger country of the USA, where Taylor and Ford had revolutionized methods of production and Sears Roebuck methods of marketing. Loewy brought with him to America a European sensibility, formed by the mixture of new industry and traditional architecture in the Paris of his youth, but had no qualms about putting it to the service of American values. Unlike many others, who followed him across the Atlantic from Europe in later years, Loewy brought little intellectual baggage, just his aesthetic sense and his gifts as a communicator and visualizer: for some of his later critics, arriving with only genius to declare was not permitted.

4935

from Zenith to Nadir

'In America, industry wants design counsel from the designer which will enable him to make more money through greater sales. Pride, social consciousness, and the desire to serve mankind better, don't seem to enter the industrialist's head.'

F.E. Brill, General Plastics Inc, New York, 1934

In 1930 the novelist Sinclair Lewis became the first American to win the Nobel Prize for literature. At the same time, Raymond Loewy was busy trying to sell industrial design to the Babbitts of Middle America. Sinclair's 1923 novel *Babbitt* charts the rise, wobble and rise of a Midwestern realtor, blustering and opinionated, hypocritical and gullible, a staunch defender – a Booster – of the American way, whatever that may be. As a comic character, Babbitt is a subtle parody, bursting with contradictions. 'Here's to the Standardized American Citizen.. fellows with hair on their chests and smiles in their eyes and adding-machines in their offices!' declares the hero in an address to the Zenith Real Estate Board, 'The modern American business man knows how to make it good and plenty clear that he intends to run the works. He doesn't have to call in some highbrow hired man to answer the crooked critics of the sane and efficient life.' Loewy described

Raymond Loewy standing in front of a restored GG-1 locomotive.

Babbittland versus Babel: the suburban calm of Forest Hills *(facing)* compared to Times Square at night *(above)* or downtown Boston *(right)*.

the members of such Booster's Clubs as 'rough, antagonistic, often resentful'. 'My French accent was no help', he adds: as Babbitt 'entertained a Smoker of the Men's Club of the Chatham Road Presbyterian Church with Irish, Jewish, and Chinese dialect stories', it is possible to see why. 'I remember', writes Loewy, 'long gloomy winter nights in Cicero, Illinois, feeling sick, tired and discouraged trying unsuccessfully to sell industrial design to minor, but tough, Midwestern manufacturers. Dismal memories: cold beds, cold meals, cold rain.' Babbitt, of course, a few years before, has an unshakeable faith in his town. 'Everyone knows Zenith manufactures more condensed milk and evaporated cream, more paper boxes and more lighting-fixtures, than any other city in the United States, if not in the world. But it is not so universally known that we also stand second in the manufacture of package butter, and somewhere about third in cheese, leather findings, tar-roofings, breakfast food and overalls!' But the faith is roundly walled with narrowness and cant ('there's a whole lot of valuable time lost even at the University studying poetry and French and subjects that never brought in anybody a cent.')

If such were Loewy's daily contacts at the start of his business, travelling endlessly across America looking for customers, his stamina, let alone his enthusiasm, must have been considerable. Yet, according to Elizabeth Reese, writing in the recent *Loewy* catalogue, he never lost his 'relish for the American character, its brashness, humour, daring, impulsiveness, enthusiasm', even if he equally carefully nurtured an independent Gallic aura for himself. His first employee in the new design office, in 1934, was an American, A. Baker Barnhart, in whom Loewy felt he had found 'the real American thing'. As customers began to come in, 'Barney' Barnhart's role in dealing with them,

Raymond Loewy, William Snaith, Jean Bienfait, A. Baker Barnhardt and John Breen, the key members of Loewy's 1930s team, in a photo taken in 1949.

in their own terms, became most important.

Loewy's first customers were described by him in a survey of ten designers conducted by *Fortune* magazine in 1933: he had designed a duplicator for Gestetner, an automobile for the Hupp Motor Car Co (it went into production only in the following year), and was working on kitchen equipment and bathroom units for Sears Roebuck. He correctly said he was a staff of one, though he also claimed he had been an industrial designer for six years, and – perhaps pardonably in such a magazine – overstated his fees, giving 3,000 dollars up as a flat fee, and 10,000 to 60,000 as a retainer. This put him in the middle rank of the other main designers: Walter Dorwin Teague, with a staff of six, claimed the same six years' experience and flat fees of 500 to 10,000 dollars, with retainers starting at 12,000. Norman Bel Geddes, with 30 staff, unhelpfully gives fees plus royalties of one to one hundred thousand dollars. Harold

The other three designers: Norman Bel Geddes in his office with Harold Van Doren (on Bel Geddes's right) *(left)*, Walter Dorwin Teague *(facing, left)* and Henry Dreyfuss *(facing, right)*. Bel Geddes's *Airliner No 4* is illustrated below.

Van Doren, who lists 'ghost writer' in his previous experience, mentions a consultation rate of one hundred dollars a day.

The previous experience of these designers is important: hardly any of them had been working as industrial designers a decade earlier. The Great Depression of 1929 may not have created the profession of industrial designer, but it sharpened and accelerated a trend towards the employment of designers in American industry. This can be seen as a worthy process – the recent *Graphic Design in America* catalogue speaks of the positive effect of the Depression 'maturing' industrial design 'into a profession with direct economic and social value'. Or it can be seen as purely economic: 'industry is only interested in those designs that there are good business reasons to manufacture' in the words of Vaughn Flannery of Young and Rubicam, in the same 1934 questionnaire that prompted F.E. Brill's comments at the head of this chapter. Loewy endorsed much of this robust approach, writing in *Industrial Design* that 'success finally came when we were able to convince some creative men that good appearance was a salable commodity, that it often cut costs, enhanced a product's prestige, raised corporate profits, benefitted the customer and increased employment.'

Many of the first generation of American industrial designers, as has often been noted, came from backgrounds in illustration (as Loewy, who also described himself in *Fortune* as having being an electrical engineer) or from the theatre (Norman Bel Geddes and Henry Dreyfuss) or from advertising (Walter Dorwin Teague and John Vassos). Only George Sakier admitted to having previously been a mechanical engineer. This background is taken by some to explain, if not excuse, the superficial nature of some industrial design of this period – styling, not design. But a wider view is possible. In the first place, many of the items on which designers were working – radios, refrigerators, cameras, vacuum cleaners, motor cars – had hardly existed as commercial mass products a couple of decades earlier. A vocabulary of form for them was an immediate necessity, if such products were to function in the home and in the marketplace: the contemporary desire for products to look modern and up-to-date does seem to have been a real desire, however it may have been promoted and exploited, even, by advertisers and manufacturers. Furthermore, many industrialists were agreed that if design was to have a real role to play, it could not be a superficial one. In the 1934 survey mentioned above, for example, the question was put: what does the designer need to know of the methods and problems of production? 'As much as possible', replied Lord & Thomas, the New York advertising agency, a reply echoed by almost all the respondents. Frederick J. Wolfe of Anglo-American Oil put the answer the other way: 'the commonest weakness of the average designer is his lack of understanding of the object he is required to design, and of the costs involved in production.' The will to move design beyond styling seems to have been present, but the immediate task in the aftermath of the October 29th Wall Street crash was to get the economy moving again. Manufacturers needed new ideas and techniques to bring their products to the public, and the designers – formerly in illustration, the theatre and advertising – were lining up outside their doors ready to show them how.

One door on which Loewy had been calling for over two years was Sears Roebuck, one of the largest American mail order companies. In 1934, Sears Roebuck agreed to let Loewy redesign the 1934 Coldspot refrigerator. It was to be the job that made Loewy's immediate reputation, and began to bring the clients to his door, in their turn. The design problem on the Coldspot was much like the one with Gestetner: a perfectly functional machine, but that looked like a machine – you might need one, but you would hardly want one. Loewy and his team quickly found that drawings did not help in trying to work out a redesign so, as with the Gestetner machine, they approached the task through models. A wooden block model was built, slightly smaller than the full machine, coated with modelling clay , and the design was worked out in the round. There were some problems with this approach: the clay was more workable when heated, so a 'clay warming oven' was installed in the studio. The first block models – or 'bucks' – were almost too heavy to move without damage, and had to be rebuilt on low bases fitted with rollers. The finished designs were painted and lacquered with care: even the wooden handles were

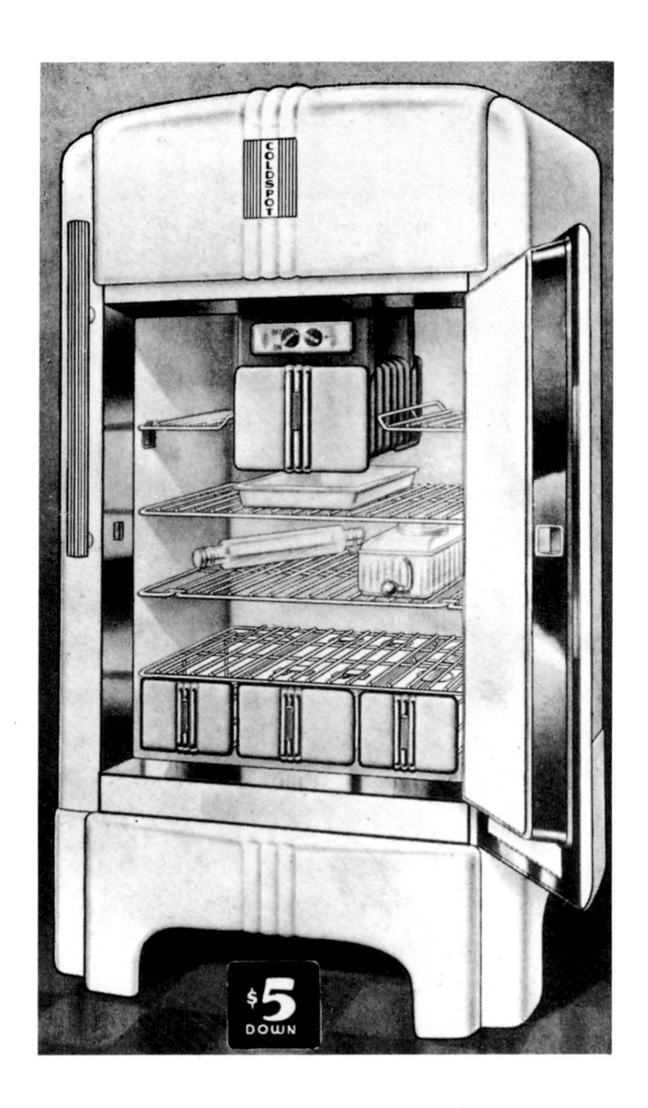

coated with copper paint which was then chrome-plated to give as realistic a finish as possible.

The existing Coldspot design was boxy, the doors divided into panels, and with wasted

Advertising *(facing)* showed the interior of the redesigned Coldspot: later Loewy designed for Frigidaire (right).

space under the machine. By moving the position of the main motor and pump, as well as remodelling the casing to have large, uninterrupted surfaces with curved corners and rolled edges, Loewy produced a refrigerator that looked immediately modern, with white easy-to-clean surfaces that suggested freshness and

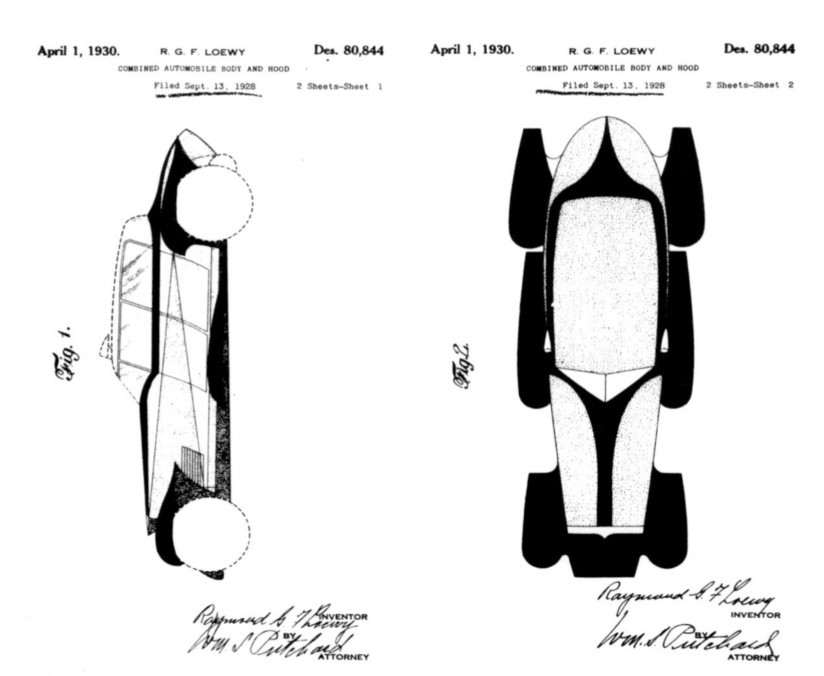

hygiene. Later models filled the space under the fridge with a storage shelf. Other innovations included replacing the interior metal wire shelving, that was costly to manufacture and prone to rust, with perforated aluminium shelves – Loewy had been working with aluminium for car radiator grilles. In addition, the latch became a long vertical bar that could be opened by a busy housewife, both hands full, with a touch of the elbow. It was also chrome-plated – 'substantial as well as attractive, like the door handle of an expensive automobile'.

The new design resulted in an immediate increase in sales: Loewy, who had received a $2,500 fee for the first design, was commissioned to design the 1935 model for triple that fee, as sales climbed to 65,000: when they

Loewy's patent application *(facing)* came nearer reality in his designs for the luxury 1932 Hupmobile Spyder Cabriolet *(above)*.

reached 275,000 the fee went to twenty-five thousand dollars. Sears Roebuck featured the design in their catalogue descriptions, and their salesmen reported the appearance of the Coldspot – also advertised at half the price of a comparable fridge – as a main factor in increasing sales. The demand for kitchen appliances and labour-saving devices at this time was certainly growing, which contributed to the surge in sales. But Loewy's intuition about the right shape this time served him well. When he came to write *Never Leave Well Enough Alone* in 1950, one chapter contains a fly-on-the-wall view of the progress of a new design for a refrigerator. The fictitious client is the Nadir Freezer Company, a subtle acknowledgement that with the Coldspot Loewy's fortunes began to rise: the Babbitts were buying..

It was John Mitchell, a friend in the Lennen and Mitchell advertising agency that introduced Raymond Loewy to the Hupp Motor Car Corporation. From childhood Loewy had sketched and drawn cars, trains and planes, and his advertising illustrations had included drawings of cars, for Pierce Arrow and Buick. In 1928 he had filed for a patent – awarded on the 1st of April 1930 – for a 'combined automobile body and hood'. The drawing accompanying the patent shows a sleek shape, with split windscreen, teardrop shaped mudguards and run-

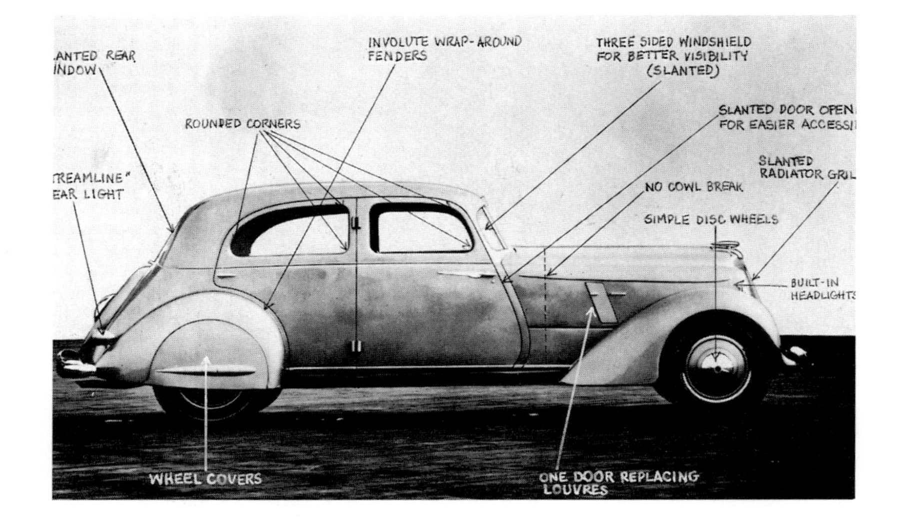

ning boards, and a curved tail. It was, in 1928, a considerable departure from the boxy horseless carriages some manufacturers were still producing, but by 1930 there were several cars actually on the road with a similar look. The Hupp brothers first asked Loewy, in 1931, about the styling of the 1932 model, the V8 Spyder cabriolet, a two-door sports model. Loewy made a number of suggestions: from surviving drawings the resulting car would have had a very even, almost classical profile, and considerable grace. According to Loewy, the management 'resisted' his new ideas, though it may simply have been that there was no time to make radical changes before the model was launched. Loewy, to prove the validity of his ideas, customized a standard car to his own design and at his own expense – something between eight and twenty thousand dollars – and toured it in France, where it won a number of motoring awards. The management, impressed, invited him to design the 1934 model, the Hupmobile, a four-door saloon.

The new aerodynamic features of the 1934 Hupmobile are highlighted on this annotated photograph *(above)*: two door *(facing)* and four door models were proposed.

This design, which was put into production in a modified form, marked a considerably more radical development than the earlier custom model. In the first place, the body was conceived as a single unit, with headlamps and

mudguards integrated into the whole, an unbroken line to the bonnet, angled rear window and rounded corners to the side windows. The windscreen was split into three sections, the radiator grille slanted and the front bumper was curved around the mudguards. A two-door model with dickie seat carried the rounded, flowing look even further. The Hupmobile flying letter H – a logo reminiscent of a certain brand of vacuum cleaner – had been replaced by a discreet badge.

There is a clear comparison to be made with another car, the Chrysler Airflow, launched in the same year. It shared an integrated and rounded styling to the bodywork. The Chrysler is more radical: the radiators, headlights and front mudguards are unified into a

Engineer meets artist: Cal Breer's Chrysler Airflow *(facing)* may have looked similar to the Hupmobile *(above)* but the conception was very different – and neither car was a market success.

single curved whole, for example. Carl Breer, who designed the Airflow with Chrysler's engineering staff, based his design on wind-tunnel tests, after modifying the traditional layout of engine, chassis and passenger compartment to provide a smoother ride. The engine was moved forward to sit between, rather than behind, the front wheels, and the rear wheels moved back to behind the passenger compartment. The frame construction of the car body surrounded the passengers, as in a modern car: another considerable innovation.

Outwardly, there is much to compare: but the Chrysler combined innovations in the engineering of the car with innovations in styling, both under the same team. The Hupmobile was original in its design, but mechanically conven-

tional. Loewy gave the car a new look because he felt cars should be that way; Carl Breer worked from theories about air resistance, lift and drag (Orville Wright was a consultant on the wind tunnel tests). Loewy seems to avoid using the word streamline in writing about the Hupmobile, except about the rear light ('streamlining,' he once said to a reporter, 'is a

Trashcan specialist and railroad baron: Loewy with Clements outside the PRR office *(above)* and one of the many fanciful designs he showed his client *(facing)*.

state of mind'). He prefers the nautical metaphor 'tumblehome' to describe the tapering of the upper bodywork.

The Hupmobile did not sell very well, and the model was withdrawn in 1936. The company disappeared as an independent producer in 1941. The pressure from the large Detroit manufacturers and from Ford, as well as the 1937 recession, probably contributed to this. The Chrysler Airflow was also taken off the market because of poor sales, in 1937. Henry Dreyfuss described the Airflow as a design ahead of its time: certainly by the end of the 1930s many of the stylistic features – integrated headlamps, rounded lines to trunk and radiator – of the Hupmobile and the Airflow were to be found on mass-market cars.

From the description in his biography *Never Leave Well Enough Alone*, Loewy did not enjoy his first brush with motor manufacturers, who spent much of their time telling him what he could not do, or ignoring him completely. The next time he worked with a car maker, Studebaker, he would organise things differently. But the invitation from Studebaker did not come until 1938, and in the meantime Loewy got busy on another form of transport: the railway train.

Having been in the doldrums after the First World War, the American railway business be-

gan to get increasingly competitive, bcth internally and under threat from the growing availability of private cars and the development of passenger air services. The railroad companies strove to offer their customers new standards of comfort and service and, above all, modernity and efficiency. Designers like Otto Kuehler and Norman Bel Geddes had published designs for streamlined locomotives and trains at the beginning of the 1930s. Theodore Clement, the president of the Pennsylvania Rail Road, then the largest network in the USA, at first tried to brush off Loewy's approach. Loewy's persistence was rewarded with a commission to redesign the trash cans at the New York Pennsylvania Station. Several days on-site research produced a new solution. Clement approved it, and at his next meeting with Loewy ('How's the great trash can specialist today?') showed him the design for a new electric locomotive, the GG-1: could Loewy do anything with it?

Loewy's solution was twofold. He cleaned up the lines of the design, emphasizing the

Streamlined locomotive designs by Henry Dreyfuss *(facing)* for the New York Central and by Otto Kuehler for the Gulf, Mobile and Northern Railroad.

robust and powerful overall shape with a pattern of five curving gold lines that swoop down to the ends of the locomotive. He went further, however, in suggesting that the railmen borrow from car manufacturers the idea of separating the body shell from the mechanism within, and producing it from welded, rather than riveted, sheets. Not only would there be costs saved, but the outside finish of the locomotives would be much smoother. The engine had a central cab and a 0-4-6-6-4-0 layout, to operate in either direction, and the new shape was a symmetrical, semi-streamlined statement of considerable power. Clement and his chief engineer, Fred Hankins, approved the design, and over 50 GG series locomotives were built. Loewy was continually involved in the design process, going down to the sheds to see the engines being built, and on one occasion marking up modifications on the full-scale model by clambering over it, using white adhesive tape and coloured chalks.

The most important commissions, for

The gold-striped GG 1 engine, after a recent restoration, *(above)* and *(below)* the bar lounge of the Broadway Limited, also designed by Loewy for the PRR.

Loewy, from PRR were those for the design of steam locomotives, that followed the GG-1. The first of these was a special version of the standard K4 Pacific locomotive. Loewy had suggested a more aerodynamic form to Fred Hankins: other railroads were already using

Me and my train...Loewy in front of the SS1, his first steam design for PRR (facing).

Designs for the SS1 *(above)* and the finished locomotive *(below)*.

streamlined locomotives – for example the New York Central Railroad's *Commodore Vanderbilt*, and the Pennsylvania Railroad would need to be seen to be up there with the competition. After a number of tests from the footplate by Loewy himself, in tweed cap and goggles, during which he remarked the absence of toilet facilities in the driving cab, further tests were made on a number of shapes in a

Loewy's third locomotive for PRR, the T4 designed in 1942: *(below)* a contemporary newsreel shot of the three locomotives together: Loewy has left a vivid account of this.

wind tunnel at New York University. Previous streamlined engines – the *Commodore Vanderbilt* and GG-1 among them – had used a shroud form, enveloping the whole engine. Loewy wanted to use the torpedo shape for K4-S, with a pointed cylinder over the boiler and a curved skirt over the front and wheels, resembling a torpedo resting in its cradle. In *Locomotive* he claimed that the wind tunnel tests had shown

the shroud form less effective at deflecting smoke from the driving cab, so reducing visibility.

The K4-S was renumbered Engine 3768, and went into service on March 3rd, 1936. 'It was an even greater thrill than I had expected', wrote Loewy. It was painted a dark bronze colour, with silver and gold lettering: it pulled the *Broadway Limited*, for which Loewy designed the interiors, between Chicago and Fort Wayne, while the GG-1s pulled the *Congressional Limited* service between New York and Washington. The profile of locomotive and tender was simply dramatic, and in motion the highlighted strips and rails on the front con-

A childhood dream: the Lionel model of the S-1 locomotive.

veyed an immediate expression of speed. Loewy was often photographed in front of or standing on this engine, which was clearly for him a very major design, crowning the second decade of his life in America. And, as he unashamedly says in the preface to *Locomotive*, fulfilling a childhood dream. It fulfilled other childhood dreams when the Lionel toy company issued an O-gauge electric model of the locomotive and tender.

Loewy also designed the T-1 engine, intended for long-haul heavy passenger work. The T-1 is best described as Rugged Moderne: a semi-streamlined boiler sits on a wide bar, placed deliberately high and cut away below to

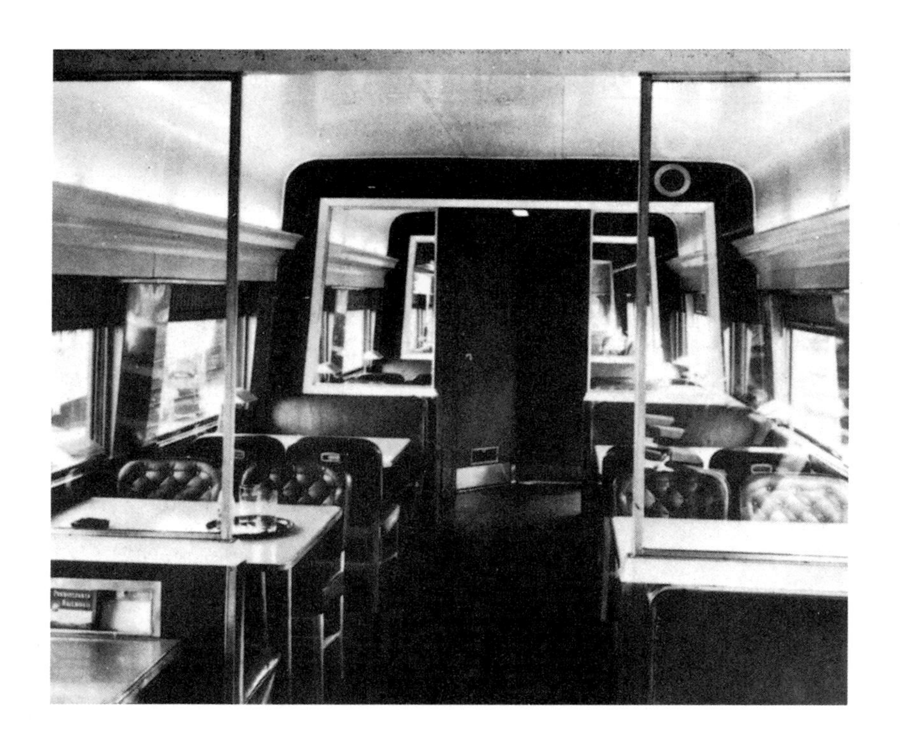

show the double set of pistons, wheels and rods to most powerful advantage. One the basis of engine 3768, Loewy designed the S-1 in 1937, an equally stunning piece of streamlining. His own description of the train coming into sight down the straight track near Fort Wayne, Indiana, often used for speed trials, is well-known: 'It flashed by like a steel thunderbolt, the ground shaking under me, in a blast of air that almost sucked me into its whirlwind. Approximately a million pounds of locomotive were crashing through near me. I felt shaken and overwhelmed by an unforgettable feeling of power, by a sense of pride at the sight of what I had helped to create...'

Interior for the Broadway Limited by Loewy *(facing)* uses Moderne elements, while Paul Cret's interiors for the Santa Fe line borrow Pueblo Indian motifs.

Work on locomotive designs led to other work for the railroad and its associated companies. One of these operated a ferry service is Chesapeake Bay, and Loewy was called on to restyle one of their ships, rechristened the *Princess Anne*, and launched in 1936. The bold curving superstructure, and the striking white and blue colour scheme, made a handsome and unusual vessel. This cheerful effect, and a new layout and decor to the passenger lounges with a dance floor, snack bar and restaurant, made a ferry trip into a pleasure cruise. Loewy summarised his work for Pennsylvania Railroad as doing 'the colour of a ferryboat, the design of a menu, a new signal tower, a bridge over the

The Princess Anne ferryboat: turning a crossing into a cruise.

Potomac, a coffee cup, or the design of a bronze tablet for a retiring executive, and.... complete blue ribbon trains such as *Spirit of St Louis*, *The Admiral* and the *Broadway Limited*. We even designed the toothpicks.'

Other designers were working for the railways as well. Henry Dreyfuss designed the locomotive and interiors for the New York Central's *Twentieth Century Limited*. He also adopted a torpedo design for the locomotive, and a restrained Moderne style for the carriages. Paul Cret, otherwise head of the University of Pennsylvania's School of Architecture, designed the fittings for the Santa Fe Railroad's *Superchief*: as the line ran through the territory of the Pueblo Indians, their tribal motifs were used on the textiles and wall decorations.

Streamlined trains continued to be popular well into the 1940s, and it was their popularity that led to their still being built. Adding streamlined cladding to a standard locomotive and tender could add five tons to the weight, before any savings in aerodynamic efficiency

were earned. But the public wanted the new trains – just as it wanted new-looking cars, and modern-looking fridges, and this the designers could provide. In the 1930s, the industrial designers had shown manufacturers that they could deliver what the public would buy.

The decade that had brought Loewy to prominence was one of considerable economic and social changes in America, and industrial design was part of those changes, though not an agent of change. For Loewy, his first twenty years in America must have seemed extraordinary. Getting off the boat with nothing, he had created a first career for himself as a fashion illustrator – he claimed to have made 40,000 dollars a year. His savings were wiped out in the Wall Street Crash, and he started a new career: ten years later he had over a dozen major corporations – including Coca Cola, Studebaker, Penn Railroad, Frigidaire – as clients, and offices in New York, Chicago and London.

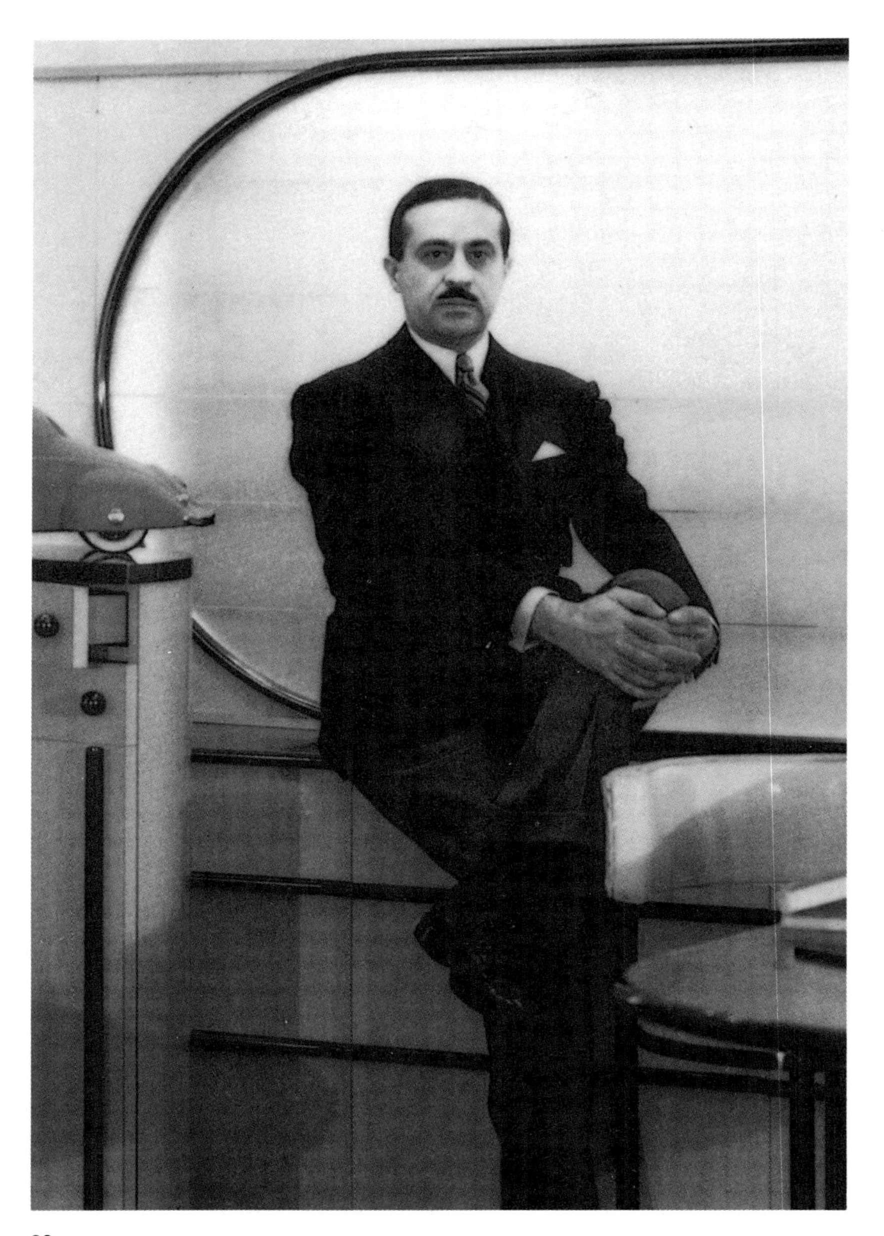

rocketports and starliners

'It was the paradox of all paradoxes... It was good, it was bad; it was the acme of all crazy vulgarity, it was the pinnacle of crazy inspiration.'

Sidney Shallett, 1940

The New York World's Fair may have claimed an international title, but by the time the Fair was to open in 1939 attention in Europe was already concentrating on more immediate and local events. Several of the foreign pavilions were finally cancelled: the Russian one, for example, was transformed into an 'American Common', complete with bandstand. The Fair rapidly became an American celebration, a reward for the difficult years after the 1929 Wall Street Crash, and a positive statement about the future. Thus the Fair looked forward from the changes that the 1930s had brought to America, and industrial designers – their own profession barely a decade old itself – played an important role in the creation and the visualization of the Fair.

The designers Dorwin Teague and Gilbert Rhode were among the informal group of one hundred businessmen who first proposed the idea of a fair to sponsors and to the US government in 1935. Teague was to remain influential in the planning stages, in particular in helping

Raymond Loewy in the Designer's Studio exhibited at the Metropolitan Museum of Art, 1937.

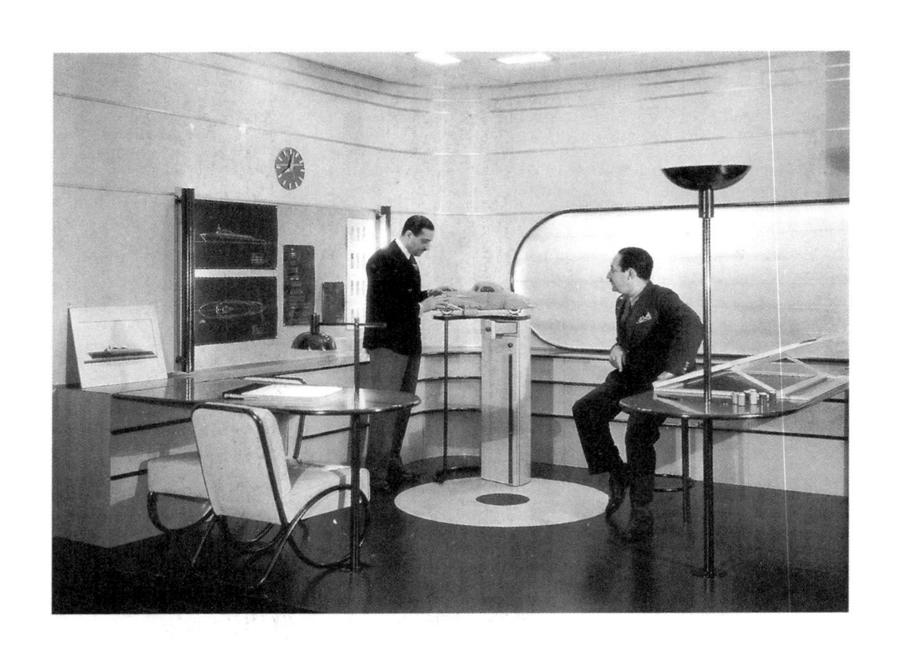

to turn the idea of the Fair away from a retrospective celebration of American history and towards a projection of the future. Such a link between industrial designers and the future was not a new one, Museums and galleries, among them the Museum of Modern Art in New York

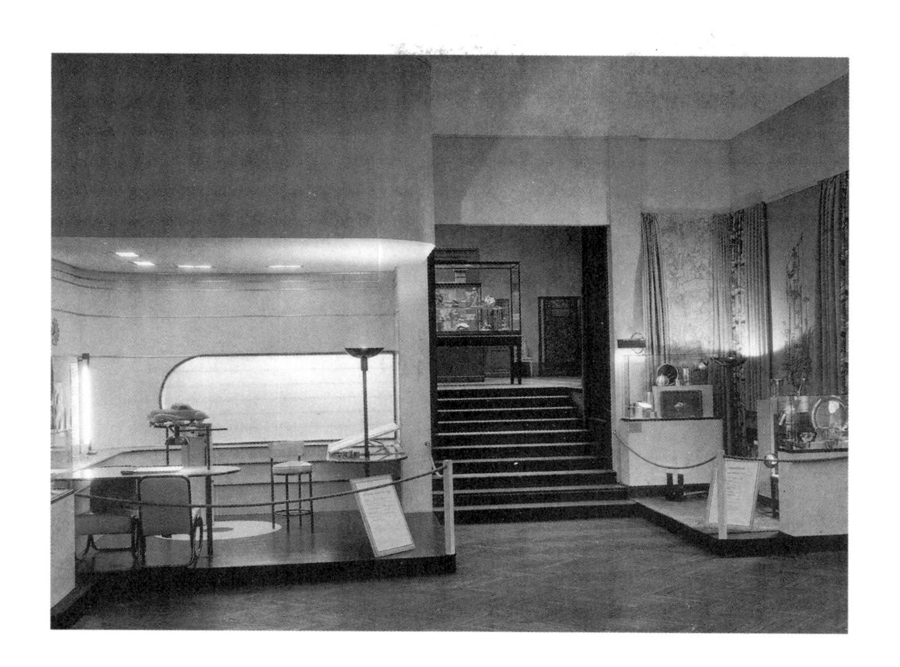

Loewy and Simonson in the Designer's Studio *(facing, above)*: note the model for the Hupmobile and the plans for the Princess Anne. The scale of the installation can be judged from the general view *(above)*. Loewy also designed a room for a child *(facing, below)* including a pedal toy in the form of an airplane.

and the Metropolitan Museum of Art in the same city had already exhibited 'industrial art' in the 1930s. The newly founded Museum of Modern Art had organized an important Machine Art exhibition in 1936 and started in 1938 a series of annual exhibitions in which members of staff selected new mass-produced objects, as examples of good modern form. The Museum's permanent design collections – and the Museum shop – continue the same tradition today. The Metropolitan Museum had started showing industrial art earlier, in 1934, and with a different approach, inviting designers to furnish complete rooms. In 1937 Raymond Loewy, with the architect Lee Simonson, had presented 'A Designer's Studio'. The walls are boldly lined with horizontal wall bars, the windows end in semicircles and a target pattern decorates the floor. The furniture – spare rods and planes in leather and metal – is appropri-

Raymond Loewy's clothes for the future, in Vogue, 1939, *(facing)* were rather more traditional than Gilbert Rhode's proposal *(right).*

ately minimal in decoration. (Loewy described the studio as 'a clinic – a place where products are examined, studied and diagnosed.') On the wall by one desk drawings for the streamlining of the Princess Anne ferryboat are displayed, while the model stand supports a study for the Hupmobile. In a contemporary photo, Raymond Loewy is seen perched on one of the wall bars, looking every inch the professional, his hair sleek, his pose alert but relaxed, and dressed in jacket and flannel trousers, casually formal. If the image recalls young movie stars, then it is not just in the clothes but also the setting: art directors like Cedric Gibbons of

Metro-Goldwyn-Mayer and Hans Drier at Paramount had been borrowing from the Bauhaus for their set designs in much the same way for years, following a trend created by European directors, such as Fritz Lang.

In the run-up to the World's Fair, other media showed their interest in a future theme. In 1938 Vogue invited nine designers, Raymond Loewy among them, to design clothes for the woman of the year 2000. Henry Dreyfuss equipped his model with a combined purse and electric fan, while Gilbert Rhode showed the man of the future in a beryllium-threaded jumpsuit, heated and cooled by the resulting 'omega waves', and with a halo-like antenna on his head for the radio and telephone attached to his belt. Raymond Loewy aimed for simplicity, giving his model a transformable gown and a simple felt bag as all the luggage the traveller of the future would need.

The New York World's Fair took its theme of Building the World of Tomorrow more seriously. The Fair exhibits comprised a mixture of buildings housing the displays of official bodies and buildings sponsored by commercial enterprises, but both echoed the general themes of the Fair. In both sectors – the private and the public – industrial designers found free play for their ideas. As befitted someone whose career had started in the theatre, Norman Bel Geddes provided one of the most dramatic and popular displays. The Futurama showed an immense model diorama of the city of 1960, with multi-lane highways linking area to area, shopping to leisure, suburb to centre, around a landscape of Moderne style skyscrapers. The subtext of this lavish display was that the automobile was the transport system of the future, and the display was to be found in the General Motors building. Bel Geddes developed the superhighway idea further in his 1940 book, *Magic Motorways*, itself based on the research done for Fu-

Henry Dreyfuss's futuristic evening gown *(facing)* for Vogue included a portable fan.

turama, and which argued the case for a series of major, multiple lane highways networking America. This was one prediction from the

Views of the World's Fair: the Mercury figure outside the Ford building *(facing)*, the National Cash Register motif *(right)* and a general view from beside the Perisphere *(above)*.

World's Fair that did come close to realization, though without the planning and ordered discipline its author had proposed – and without all the social benefits he had envisaged either. Sadly, Bel Geddes career declined in the 1940s, and he died in 1958, leaving *Magic Motorways* as part of his legacy, together with a plan to revitalize Barnum's circus using pink sawdust. His other, perhaps better known, book is *Horizons*, an unashamed essay in streamlining. It is rightly the streamliner's bible: his single wing air-planes, teardrop ocean liners and aerodynamic cars and buses make an elegant and coherent case for streamlining as a style.

Back at the New York World's Fair, another urban vision was offered by Henry Dreyfuss with his Democracity, displayed in the Perisphere. The sphere, two hundred feet in diameter was with the Trylon one of the official landmarks of the Fair, and within it visitors

An external view of Bel Geddes's Futurama at the World's Fair *(facing)*: Dreyfuss's Democracity *(right)* added social values to motorway magic.

were led onto a balcony running around the inner circumference, from where they looked down on a model of the landscape of the future. Here too multi-lane roads linked district to district and town to town, with dramatic bridges over rivers or elevated sweeps around massive skyscrapers. The play of light suggested the passage of night and day, while above the spectators' heads a series of multiple projection screens showed clips of film about the different classes of men and women who lived and worked together in Democracity. Here, in contrast to the Futurama, the emphasis was on civil order and social progress,

Part of the model diorama in Geddes's Futurama *(above)* and Dorwin Teague's display for Ford *(below)*.

rather than the advance of the machine. As a designer, Henry Dreyfuss was well-known for the humanistic approach of his work, which would develop into full scale ergonomic studies during the 1940s. The development by his office of the John Deere tractor, in which the comfort and safety of the farmer was a predominant factor in the re-design, is a classic story, as is his work on the development of the handset for Bell Telephone – he is said to have got the contract in the first instance by refusing to pitch in a sketch for the proposed design before seeing the full brief, unlike the other agencies approached.

Another designer whose work may be said to have reached its apogee at the Worlds Fair is Hugh Ferriss, the architectural writer and illustrator. His 1929 book *The Metropolis of Tomorrow*, a title that consciouly or uncon-scioulsy recalls Fritz Lang's 1927 film, uses dramatically-lit monochrome drawings of existing buildings to pave the way for future urban landscapes. The skyscrapers and flyovers that fill these later drawings articulate a dream of order and mechanical power that has obvious parallels in Bel Geddes's work. The fact that Ferriss was appointed official artist to the Fair strengthens this link – many of the drawings of the exhibits at the Fair have all the signs of his style, with hanging perspectives and sombre chiaroscuro effects.

Walter Dorwin Teague's participation in the original planning of the Fair was perhaps acknowledged in the invitation to his office to design the interiors of the official US Government pavilion. In the State Reception room he strove to show that 'formality and dignity can be expressed in current terms of line and mass, without elaboration and pretentiousness'. Together with the architect Albert Kahn, he designed for the Ford Pavilion the Road of Tomorrow. This was a full-sized half-mile road, spiralling up four levels, along which visitors were invited to drive the latest cars from the Ford model range. He himself, in his 1940 book *Design This Day*, claimed that 'the spiral ramp of the Ford Building, with its sweeping curves

and its inward-sloping, angular piers, was a logical expression, in reinforced concrete, of a dynamic opposition of forces, and of lines expressing forces.' In even more visionary style he spoke of his objective as 'the aesthetic value of simple masses, integrated forms and rhythmic line'.

If his colleagues kept their feet – or at least their clients' wheels – on the ground, Raymond Loewy reached for the stars. The finale of his History of Transport, the Focal Theme exhibit

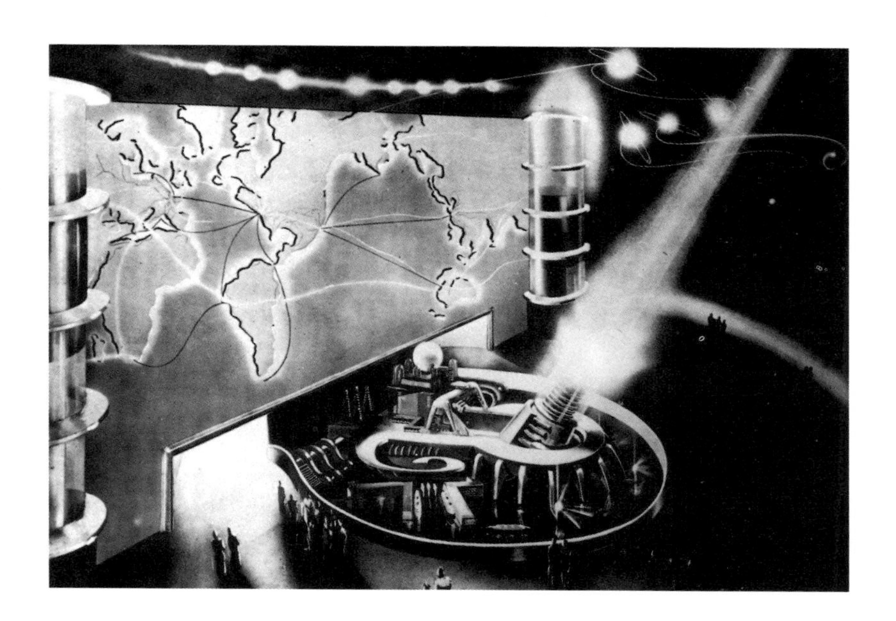

The attractions of Loewy's Chrysler 5-Star Show are shown in this leaflet *(facing, above)*: Loewy himself is seen working on the construction of the Rocketport *(facing, below)*, while an artist's impression shows the moment of lift-off *(above)*.

in the Chrysler pavilion, was the Rocketport. First came dioramas showing the development of transport, from foot to horse, from covered wagon to Zeppelin, up to the taxi, car and bus of the future, sketched by Loewy in the correct teardrop style. Then the visitor was ushered into a darkened area, where in a model display the evening New York to London passenger rocket was being readied for launching. Cars and boats disgorged passengers in the bustle, lifts took them to their staterooms, then a crane hoisted the rocket into position. Then, in Loewy's own words, 'at the sound of sirens: Lift-Off! In a moment, a blinding flash of hundreds of strobe lights and the roar of compressed air suddenly released – the rocket, through optical illusion, seemed to disappear overhead in the blackness of space. Then total silence while a gradually diminishing point of light faded away to the stars.'

The cover of *Never Leave Well Enough Alone*

Loewy's claim that his 'realistic' model represented Chrysler's 'technological leadership' is no more misguided than Teague's panegyric to his spiral ramp: anyone who has tried to drive out of an underground car park will have their own view of spiral ramps and the 'aesthetic of simple masses'. Indeed it is difficult to look back at the ideas some designers expressed at the time and not find them wholly naive and uncritical, with hindsight. But the 1930s was a decade which had reinforced America's belief in the potential of science and technology, and industrial design was one means that brought the fruits of technology to the public's notice. That conspicuous position, and the hyperbolic atmosphere of the World's Fair, may have led some designers to take themselves too seriously: Loewy, for once, seems simply to have enjoyed a very special special effect, even though years later he was straight-facedly to suggest to NASA that his Rocketport work showed him as concerned with spaceflight well ahead of his time.

Other books written by designers at around this time show a similar intentness of purpose. Books and fairs both offer designers the opportunity for design without responsibility, and for going that little bit further. The tones of voice designers use in their own books can be used as a guide to their design attitudes. Bel Geddes's books *Magic Motorways* and *Horizons*, have both been mentioned. Walter Dorwin Teague's *Design This Day* is another serious work, as its title, redolent of the bible class, suggests: the argument of the book underlines Teague's first love for architecture – he eventually qualified as an architect at the age of fifty-five, an event that was especially important to him personally. First published in the USA in 1940, *Design This Day* was issued in a revised edition in the UK in 1946. Teague was concerned on the one hand with appropriate forms, tracing this back to classical canons such as the Parthenon, and on the other with the social duty of the designer. He sees the designer as a catalyst in the humane process of bringing the benefits of technology to all. The revised edition of the book ends with a rallying cry: 'Our better world will be built because men envision it, will be united without organization or compulsion to create it.' This neo-Platonic vision of designed social order is set out in measured statements of purpose, into which the real world of marketing new products and services does not intrude.

Compared to this high-minded clarity, Raymond Loewy's 1951 book, *Never Leave Well Enough Alone*, reads more like a novel. It is brisk and fun. The name-dropping and the endless mentions of Loewy products can be wearing, but Loewy drives through it all, interrupting his text to pass on a favourite recipe for a banana dessert (coyly named 'Angel's Tails'), scattering instant lessons in graphic design onto the chapter openings, dropping in infuriating one-liners ('Industrial design technique in England has not yet reached our degree of perfection as yet'.) In among the gossip there are some of insights into Loewy's own approach. For example, he is careful to defend styling: 'there are cases in which a shell or wrapper treatment is in order or justified. I believe when a given product has been reduced to its functional best and still looks disorganized and ugly, a plain simple shield, easily removeable, is aesthetically justified. On the other hand, the designer who resorts to such a device without having assured himself that there is no straightforward design solution is guilty of professional carelessness.' He also sounds out against design fads: 'one monstrosity after another: the Modernistic, the Cubistic and the Futuristic', and he goes on to say 'we saw it all happen with consternation: the interior of the Chrysler building, the Lexington Hotel, the Roxy, the Eighth Avenue cafeterias, the Paramount Theatre on Broadway, etc etc. It was shocking and a sign of ill omen. If such nightmares of vulgarity were sweeping the country, what were the odds against us, the pure boys, apostles of simplicity and restraint?'

The apostle of simplicity also tells the story of his redesign of the Lucky Strike pack. One morning in 1941 a man introducing himself as Mr Hill called on Loewy. Taking off his jacket, but retaining his hat, a battered affair decorated with fishing flies, he took out his cigarette case and lighter. Both were by Cartier, the man pointed out. Loewy was able to show that his suspenders were also designed – to measure – by Cartier; they order these things better in France. Civilities thus established, the man explained that he was George Washington Hill, the head of American Tobacco. Someone had told him that Loewy could redesign the Lucky Strike pack, though he didn't believe it. Loewy

The Lucky Strike pack: and the best design after-dinner story ever?

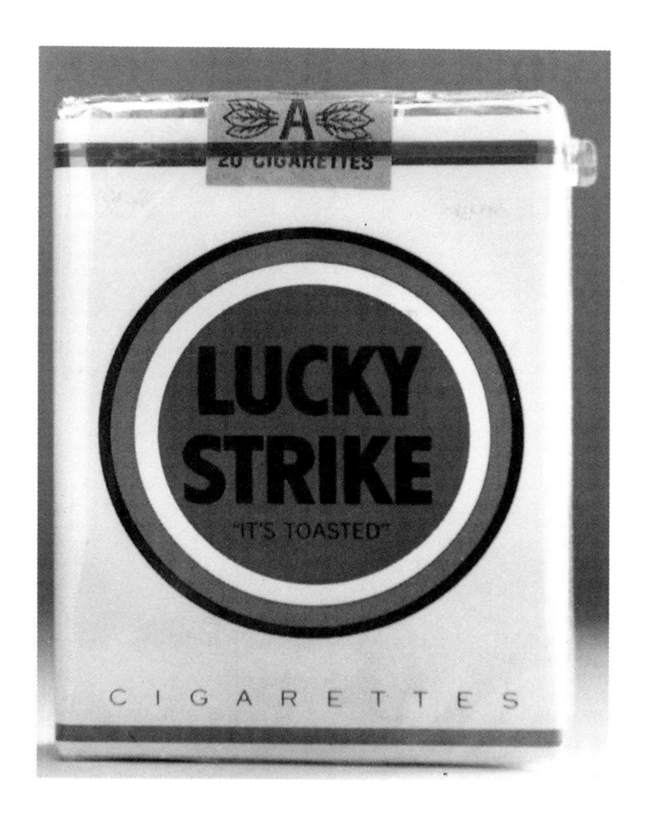

proposed a bet: fifty thousand dollars if the redesign (that would come to Loewy 'one spring morning') worked. Hill scribbled his acceptance of the bet on a visiting card, picked up his coat and left. One morning Loewy did have a go at the design, dropping the green background colour in favour of white, repeating the target design on both sides of the pack, and smartening up the typography. He won his bet.

In fact the redesign was not a whim on either punter's part: the green metallic ink used to print the old packs was not available in wartime, and research had suggested that the green colour was associated with the military. But the

story is far too good to be spoiled by such realities, and it illuminates the way Loewy liked dealing with clients: face to face, on a human level, designer to president. This is the way he did business with Gestetner, and with Clement at the Pennsylvania Rail Road. When he had to start on the ground floor, as at Hupmobile, Loewy's skills did not work so well. Loewy had been approached by Studebaker in 1938 to act as a consultant designer. Recalling the buffeting he had received at the hands of Hupmobile, Loewy proposed a radical approach. The design office would be situated in the Studebaker works, but would be a Loewy office. Studebaker executives, other than the project engineers, would be welcome in the design facility by invitation only. In the obsessively competitive world of the automobile makers secrecy over new designs was routine. Studebaker was trailing behind the big three – Ford, GM and Chrysler – in sales, and so was perhaps willing to accept Loewy on his own terms. Deliberately or not, this policy was to have one important consequence.

The outbreak of war in 1941 led to a series of governmental measures by the War Production Board to reserve scarce materials for war use

Loewy did some work for Studebaker on the pre-war President *(facing)* but the first total design, hailed as 'The New Look', was the 1947 Champion *(above)*.

(as with the metallic ink on the Lucky Strike packet), and to put industry onto a war footing. Because the conversion to war work could not be guaranteed to have even effects, and to minimize the risk of a post-war inflationary boom, a further rule forbade the use by manufacturers of human or technological resources to develop post-war models. The idea was that all sides would start out even at the end of the war. But the rule applied to in-house design facilities, owned by the companies themselves. In Studebaker's case the rule did not apply, since the facility was owned by Loewy: in fact, the 'design facility' soon consisted of only Virgil Exner and a drawing board, housed in Exner's spare bedroom, as the other designers were called up. But the result was that Studebaker was 'ready to go' on production plans for a new model the day the war ended, while other companies were left to start re-design from scratch. (Ford had run a secret design unit for part of the war, but when Henry Ford Senior got to hear of it he ordered its closure: Ford's first post-war models were to appear two years after Studebakers'.) The post-war demand for cars was enormous: three years' full production would hardly have met it, and so any new model could expect to find a market. But the break in production had a further advantage. It gave Loewy and Virgil Exner time to think out thoroughly what kind of car the post-war American public would want. Loewy had already tried to break Detroit's styling rules with the Hupmobile, and now he was going to try again. He kept insisting that 'weight is the enemy', so apart from a purer and less decorative approach to styling he also sought to make the car look lower and leaner than the bulbous pre-war productions of the 'Big Three'. The resulting car, the 1947 Studebaker Champion, was unveiled in 1946. The sales brought Studebaker back from the edge of bankruptcy, to fourth in line in the industry.

The 1947 Studebaker Champion was a radical departure. Its new features, such as the curved rear window, the understated radiator grille, and the balanced design, in which the bonnet and the boot were on the same line and much of the same length, all appealed to the public, and its unusual shape, with the front and back ends of the car parallel and roughly equal in length, with matching curved windows, provided comic writers with many gags of the 'glass backwards' variety. The car was also considerably lower, being only five foot overall, and wider than had been conventional. Arthur Pulos described the success of the Champion as 'cultural as well as financial': it set a marker for post-war car design. In addition, its success proved that industrial designers were capable of anticipating and satisfying public taste in the key mass market of the automobile. That lesson was not missed. Ford, for example, where there was an existing in-house design team, brought in an outside designer, George Walker, to remodel their 1949 model, and he adopted the cleaner lines and flatter profile typical of the Studebaker for the car: he also adorned it lavishly with chrome, encouraging the trait that

The 1950 Land Cruiser series *(above)* emphasized the absence of the radiator grille with a propeller 'spinner'.

Studebaker's publicity, as the 1950 brochure page *(overleaf)* shows) emphasized the importance of design: 'Styled ahead for years to come'; 'The New Look gives way to the Next Look...'

was to characterize Detroit cars of the 1940s and 1950s. In Loewy's words 'it has been the designer's sad experience that a great majority of the people love chrome, and love it dearly..'

Loewy's 1950 Studebaker Land Cruiser – 'a melody in metal, a symphony in steel, beautifully streamlined, excitingly new', to quote the advertising jingle – kept to the clean lines of the Champion, but was not so successful a design in sales terms: it did, however, make an incongruous appearance as a getaway car in the 1955 British film *The Ladykillers*. And with its propeller nacelle replacing the radiator grille, the car signalled the adoption of motifs from aircraft that was becoming increasingly popular in automobile design. Harley Earl, head of styling at General Motors, had taken the twin tail booms from Kelly Johnson's Lightning fighter as the inspiration for the extravagant tailfins on his Cadillacs. Similar motifs inspired by wartime aircraft and later rockets were to be found through out the 1950s on American cars,

Studebaker Champion Regal De Luxe Convertible

Studebak

Styled ahead f

From gleaming spinner straig through to flight-streamed rear fen ers, the low-swung 1950 Studebak is a symphony of power and luxu in every sweeping line.

New in wheelbase length and ove all length, the 1950 Studebaker lavishly roomy and richly appointe

Its distinctive huge sweeps window and windshield glass assu

The new 1950 Studebaker convertible interiors are finished to perfection. Nylon cord or leather seats, with simulated leather trim, resist wear and resist weather. The fully automatic top responds instantly when you want to raise or lower it.

The new 1950 Champion instrument panel is trimmed in attractive chrome. Dials easy to read. Clock at extra cost. Large package compartment. Built-in ash tray. Handsome dash panel accommodates specially engineered Philco radio —available at added cost if desired. Regal De Luxe steering wheel illustrated.

Exceptional knee-room in the front com partment is one of the distinctions Studebaker's modern designing. Sedar comfortably accommodate six passenger

Studebaker Commander Regal De Luxe 4-Door Sedan

for 1950

ears to come...

ou the extra safety of extra vision. It's new in handling ease and riding ase—with a new Studebaker-designed elf-stabilizing coil-spring suspension p front, improved symmetrically entered Studebaker variable ratio teering, extra low-pressure tires—and balance of weight that assures road-ugging sure-footedness mile after ile on straightaway or curves.

Luxurious nylon upholstery—introduced into motoring by Studebaker—is standard in all 1950 Regal De Luxe Commanders. It's easily washable and long wearing—sheds dirt instead of absorbing it. Simulated leather door panels, high-style appointments.

udebaker master craftsmen—many of em father and son teams—excel in pains-king workmanship—build surviving undness into every 1950 Studebaker.

The Commander instrument panel has big, sweep-type aircraft dials. Electric clock standard on Land Cruiser—extra cost on other models. Three-spoke steering wheel standard on Land Cruiser and Regal De Luxe Commanders. Philco radio—at extra cost—specially engineered to fit Studebaker acoustic qualities.

'Loving chrome dearly' on a 1950s Cadillac *(left)*, as compared to the 1950 Studebakers *(facing)*. Aircraft motifs now became commonplace on American cars – the bonnet feature *(below)* is from a 1957 Chevrolet.

a band-waggoning trend that recalls the inappropriate use of streamlined motifs on static objects two decades before.

In 1953, however, Loewy was to score another success. He had regularly customized cars for his own use, as with his Hupmobile, and also, for example, a Lincoln Continental – it was a habit that would continue with sports

cars including a Jaguar, a BMW and a Lancia in later years – and now, so one story has it, he did so again. He had a two-seater touring body built onto a Studebaker chassis. When he unveiled it to Studebaker management, they bought it on sight. The Starliner, as it was called on its launch in 1953 (it was also called the Fairlane), was bought as fast as it could be made, and it launched a vogue for two seater sports cars that led in short order to the Chevrolet Corvette and the Ford T-Bird. A more likely version of the story is that the idea of developing a Studebaker sports car emerged from conversations between Loewy and Paul Hoffmann of Studebaker. The project agreed, engineering and design staff brought their own enthusiasm for the idea to bear: in *Industrial Design* Loewy mentions in particular the engineers Gene Hardig and John Churchill and the designer Bob Bourke. The Starliner was first modelled as a show car in 1951 and went

on general sale two years later. According to Loewy, the car itself was a winner, accounting for over forty percent of Studebaker's sales turnover in its first year. In fact, the vehicle was underpowered, and its handling was clumsy. Harold Vance, president of Studebaker, was not in favour of sports cars, and so the whole car was not pushed further. Jay Doblin suggests that if it had been fitted with a more powerful engine and robuster handling, it would have achieved the same success as the Thunderbird. In his words, the car was 'probably Loewy's greatest design, but a marketing failure despite exceptional styling'. Many en-

Loewy at the wheel of a Starliner: taking on Detroit and winning.

thusiasts bought Starliners and fitted more powerful engines: the so-called Studillac was a Cadillac powered version, much favoured by hot rodders.

The Starliner design was notably free of chrome, and had a sloping nose and concealed radiator, features that recall European rather than American car design: Bill Mitchell of the General Motors design team is said to have learned from the car the lesson that the absence of chrome could still sell. *Fortune* magazine hailed the Starliner as the first American sports car. The *New York News* spoke of Loewy taking on Detroit and winning. From his own account, Loewy felt the competition from Detroit keenly, and the success of these two designs must have pleased him particularly. In *Industrial Design* he makes this point by referring to the side panels, where a recessed or intaglio effect on the doors gives 'character lines'. This would normally have been achieved, Loewy says, by an additional panel and trim – by increasing weight – rather than by moulding the existing panel: Loewy thought his solution was an innovation. Ironically, it is one of the features of the design that has least pleased the critics, a reaction echoed later when similar features were used on the doors of the Triumph TR7 sports car: 'He didn't do that on the other side as well, surely?' was one designer's acid comment on the TR7 design!

The engineer Gene Hardig, who had worked with Loewy on the Starliner and the Champion, also worked with him on a third design for Studebaker. This was for the Avanti, a two-door sports car introduced to the market in 1962. The design process took an astonishing seven weeks from design briefing to approval of the finished full-scale model. Studebaker wanted a two-door, four-seater car for the expensive end of the market, to be built on a standard chassis. Loewy insisted that it have a 'European feel', with stiffer suspension and more positive steering than was common on Detroit cars. He also insisted on a minimum of chrome trim, as before, on the absence of a visible radiator and a wedge shape to the front of the car, with a pinched waistline, intended to recall the Jaguars and Ferraris then racing at Le Mans. The result was a handsome design, with a distinctive forward profile and off-centre blister on the bonnet. The design was developed by Loewy, with Johnny Ebstein and two other designers, in a rented shack near

Loewy and Virgil Exner with the design team at South Bend.

Loewy's Palm Springs house. Loewy had worked up a series of preliminary sketches, and work began directly on a quarter size clay model, using the sketches as a guideline. Once Sherwood Egbert, president of Studebaker, had flown down to the improvised design studio and approved the design, the team moved to South Bend, where a full scale model was made: Loewy describes finishing some of the contours himself, using touch rather than sight to

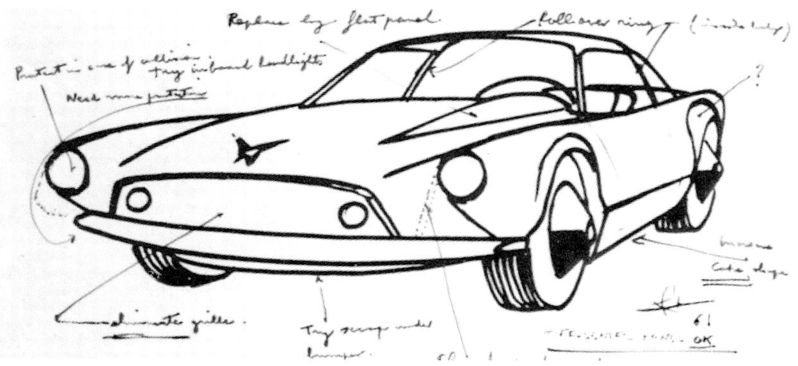

get the lines right. 'Subtle accents could not be explained any other way.'

At the same time, Ford commissioned a new design for an inexpensive sports car, to be based also on an existing chassis, but with the seat position moved back to allow a longer bonnet. Three separate design teams were put onto the project, and the final design was put

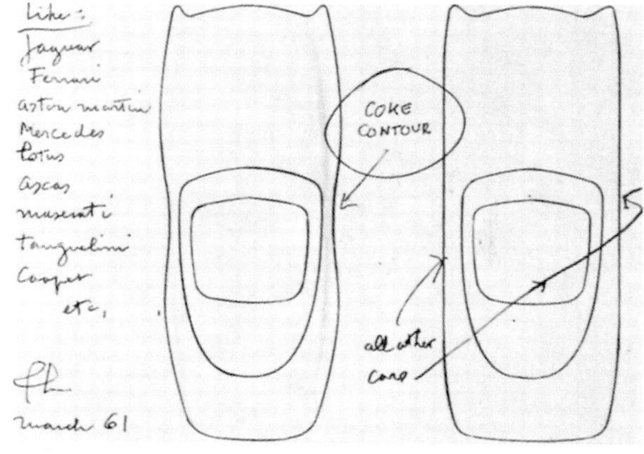

The Studebaker Avanti, seen in a publicity drawing *(facing, above)*, in a drawing by Loewy *(above)* and in two design sketches: Loewy used two Avantis, one in France, one in the USA.

into production only in 1964. But then it was the Ford Mustang, an immensely successful car in sales terms, and which captured the popular imagination in the same way that the Thunderbird and Corvette had a decade earlier. The differences in approach between Ford and Studebaker are obvious, and both approaches led to appropriate designs for their different markets. What both designs emphasize is the success designers were having in anticipating the taste of the American public, and the willingness of industry to trust designer's judgement. An article in the *New York Times* In Business pages in October 1963 reviewed the work of the six automobile designers who were responsible for the seven million 1964-model cars. The article leads with Gene Bordinat of Ford: 'he likes clean uncluttered lines of today's cars, as opposed to chrome and gingerbread of the recent past'. He also predicts that the car-buying market will become more segmented, leading to a greater variety of vehicles and models. The head of styling at Chrysler, Elwood P. Eagel, previously head of international styling at Ford, collects Oriental art, and decorates his office with it. Richard A. Teague, of American Motors, is at 39 the youngest designer featured. He had worked on aircraft design during the war for Northrop, and also believed the designer had to 'keep it simple and clean. Proportion and form are the most important factors in designing.' Whether William Mitchell, vice-president in charge of styling at General Motors ever since he had succeeded his mentor, Harley Earl, in 1958, shared the same simplifying view is doubtful: he was the oldest of the designers, bar Loewy, who is described as 'the best-known industrial designer of his time'. Brooks Stevens, the designer of the Studebaker Gran Turismo for 1964, is the last designer mentioned. All the other designers have solid experience in designing for the automobile business, and many are long-term employees of the same company: Studebaker alone of the major manufacturers uses outsiders. Loewy in his comments on the future of car design stressed aerodynamics and product quality, as opposed to boxes on wheels. Brook Stevens claimed that his design for the Studebaker Hawk had tripled produc-

The interior of the Avanti, a car whose success provoked an even greater one, the Ford Mustang *(below)*

126 MPH

tion runs. Loewy had also claimed that Studebaker had sold every Avanti that they could build. Both claims ignore the fact that at the time Studebaker was beset by industrial problems. In fact the Avanti's fibreglass body panels proved a major headache in manufacturing, and restricted the output to a mere 3,834 cars in the first year. One comment on Studebaker's choice of using outside general designers – Brook Stevens also designed that American classic, the Roadmaster bicycle – is that Studebaker eventually failed as automobile manufacturers. More to the point is the visual comment – probably unintended – in the issue of the *New York Times* mentioned above. Over the page from the comments from the designers is one of Doyle, Dean and Berbach's witty adverts for the Volkswagen Beetle, the car – and the advertising – that was showing up all Detroit's weaknesses.

If the post-war car industry needed all the help it could get from industrial designers, so did the new airline companies. At the end of the war the major aircraft manufacturers raced to put civil airliners into production. The Boeing 707, the Douglas DC 8 and the Convair 800 were the first trans-continental jet passenger aircraft. As the performance of each aircraft was similar, and fares were controlled by federal regulations, the service element became paramount in gaining and keeping customers. Boeing invited Walter Dorwin Teague to work on the interior of the 707, and he secured an arrangement that echoed Loewy's earlier deal with Studebaker. Not only would Boeing build a complete mock-up for Teague to develop his ideas, but Boeing management would not see the design until it was completed. The very considerable cost involved, let alone Teague's half-million dollar fee, was in the end justified: Teague used a complete simulation, with engine noise and exterior views, of the flight from

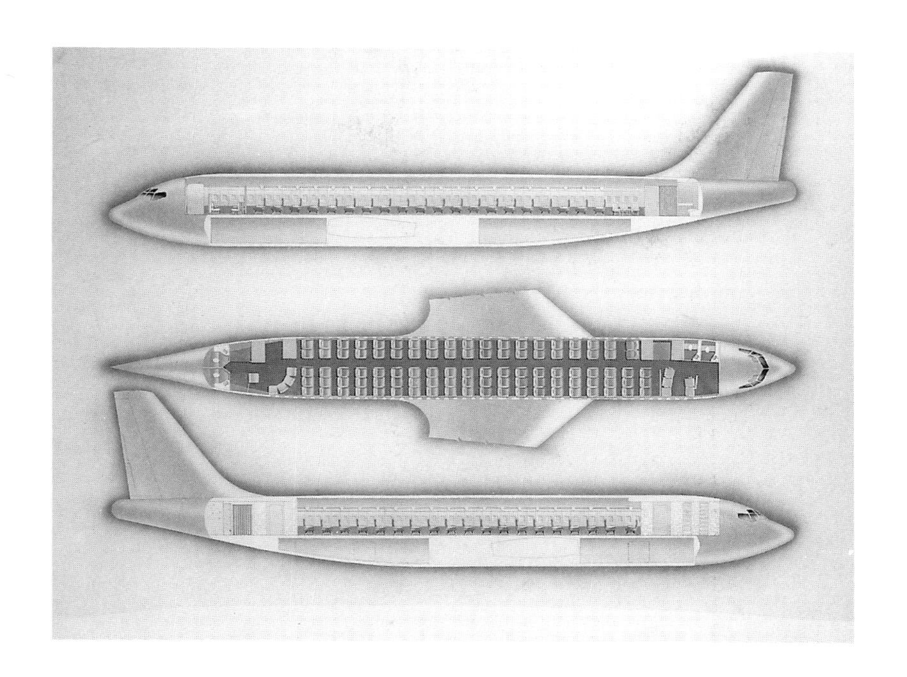

Layouts for interior seating on the Boeing 707, by Walter Dorwin Teague's office.

New York to Seattle to show his designs to Boeing management. The design, which made extensive use of newly available wallcovering materials to provide patterns and textures for the cabin interiors, was accepted at once, and Boeing built a similar mock-up themselves in Seattle to use as a sales tool in presenting the aircraft to other airlines. Teague also established a satellite office in Seattle to manage design for Boeing, while during the time the New York mock-up was in place Teague himself took great pleasure in inviting visitors to 'come and see my new airplane', to the annoyance of the Boeing engineering team who had designed the airframe.

In Loewy's case, though he had worked on the interiors of the Boeing TWA Stratoliner in 1938, his office was involved, with Douglas's in-house design teams, in developing DC-8 interiors for United Airlines. Later his Paris office was to be involved in interior designs for the Air France Concorde. He did have a hand, however, in the field of private flying. Fairchild had been in, the aviation business since the 1920s: their expertise in making wooden wings and fuselages from glued plywoods had been borrowed by Howard Hughes for the Spruce Goose, the eight-engined monster seaplane that he completed after the end of the war. For Fairchild Loewy remodeled the interior of the Fairchild F47, a low wing single engined monoplane, seating four people. There are evident borrowings from automobile interiors made, as Arthur Pulos says, 'perhaps as much for psychological as for practical reasons'.

A final transport design that should be considered is Loewy's work for the Greyhound Motorbus Company. He had been asked by them before the war to work on the corporate image, and he had instituted a flowing blue and white colour scheme and refining the running dog logo to look less like a 'fat mongrel', as Loewy called it. The World's Fair 48-seater bus by Greyhound was also designed by Loewy, though the basic model was not radically changed. He was then asked to work on the development of a new bus, to seat a larger number of passengers, and from the commission – completed before the war but not put into production until 1946 – came the Silversides motorcoach, which developed into the doubledecker Scenicruiser, launched in 1954. In 1940 Architectural Forum said of Loewy that he was 'the only designer in the United States who can cross the country in cars, buses, trains and aircraft he has designed himself'. Loewy's success in the transport field is the most spectacular part of his achievement, and the one that comes closest, in product terms, to earning him the title of 'Pioneer of American Industrial Design' used as a heading for the

Original proposal for the Greyhound Scenicruiser bus *(above)* and *(below)* the final version on a Greyhound poster.

1990 travelling exhibition about the man and his work.

4Y50·72

it's war, boys!

'It didn't take a genius to make money during the war. All you had to do was open a store and not get dead drunk. You had customers ready and willing.'

Lee Oremont in Studs Terkel *The Good War*

In 1941, as has been seen, the designers went to war along with the rest of the USA. Loewy, Teague and Dreyfuss were jointly commissioned to work on a suite of war rooms for the Joint Chiefs of Staff, and to make a set of four world globes, one each for Churchill, Roosevelt, Stalin and General Marshall. (Because aluminum was a key war material, the globes were made from hoops of cherrywood, a curiously rustic touch.) The war rooms comprised two main meeting rooms, similar in size, one equipped with wall maps and the other furnished with display systems, and a smaller adjacent room for storage. This project was completed in a short six weeks, being built to full scale in New York and then transported to Washington and re-erected.

The War Production Board was among a number of bodies established by the Federal Government not only to control the supply and use of war materials, but also to regulate industrial production, with the double aim of max-

Man on the move: Loewy with the 1939 Studebaker and the S-1 locomotive.

imizing war production and minimizing inflation during and in the aftermath of war. Industrial designers found themselves involved in two ways: in providing design services to the armed forces directly, and in helping industry adjust to the changes of wartime. Loewy, for example, was asked to redesign lipstick holders so that they did not use metal, and devised a cardboard pack, quipping that he had done something for the morale of American women, and also, he hoped, for the morale of American men! More seriously, the development and exploitation of new plastics and of glass compounds – such as Pyrex and Lucite – were encouraged by wartime pressures, and led to the peacetime acceptance of such extemporized products.

Other wartime work undertaken by the Loewy office included habitability studies for the armed forces (which led to similar commissions from the US Navy in peacetime), and projects such as the design for the Medical Air Corps of a glider that could convert on landing into a field hospital, as well as other work for the medical and engineering branches of the Army. The other designers were equally occupied: Henry Dreyfuss's office worked on the ergonomics of artillery pieces, reducing the set-up time of one weapon, the 105 mm anti-aircraft gun – from over fifteen minutes to under four minutes. One designer working with Dreyfuss at the time claimed that it was their wartime work which solidly established the human factors research for which the agency later became justly famous. Bel Geddes, who during the First World War had designed and produced a board game simulating warfare with model soldiers, tanks, guns and so on, and whose experience at building panoramas had been proved at the World's Fair, developed a series of model boats and appropriate settings, for recording sea battles for research and in-

The jeep: an American design icon, based on a failed British idea and developed by a committee.

struction, and for record purposes. These models were also used for training in aircraft and ship recognition, and in reconstructions not only for the services during and after the war – for example in recording the naval battle at Midway – but also by magazines such as *Life*. Egmont Arens worked particularly on the use of colour systems, not only for highlighting different controls and instruments or for colour-coding wiring systems (both now common practice) but also in choosing appropriate colour schemes for military environments. Other designers worked on visual and training aids, and on camouflage. Ironically, the classic product that remains one of the design icons of America's involvement in the Second World War, the Jeep, was in fact designed by a committee.

In 1943 the *New York Times Sunday Magazine* commissioned articles from Raymond Loewy and Dorwin Teague on the future of industrial design, in particular on what post-war consumers could expect by way of new products. The two arguments were set out under a dramatic illustration by Hugh Ferris, of monolithic buildings serviced by motorway ramps and heliports. The heliport was Teague's idea. He took a robust view of the future, asserting that American business has learnt to use design and innovation, and the consumer has come to expect it. 'American business is alert, flexible and adaptable', and so ready to roll out new products as soon as production can be resumed. Even, he says, 'new products will appear which will make the fanciful predictions ...of today.. seem commonplace.' His own fanciful prediction is for the development of personal helicopters as serious competition to private cars, of new petrol that will make fifty miles to the gallon a commonplace with lighter engines, and of rocket-assisted transport aircraft. He speaks of Bill Stout, former chief designer with Ford who in 1943 was working for the aircraft company Consolidated Vultee: he had designed a flying car, with folding wings, which Teague was sure would go into production. The aggressive optimism of this article contrasts with the sober enthusiasms of his earlier book, *Design This Day*.

Raymond Loewy's contribution was surprisingly downbeat. He warns against taking some published ideas about future products at face value, for they may never come about. He stresses the difficulties of retooling even when industry can step down from a war footing, and the time delays in bringing any new ideas into production, though asserting, in self-contradiction, that wartime innovations would be applied at once to post-war products. Even if someone somewhere has invented a new plastic as good as glass, he suggests, think about the time and the distance 'from the laboratory to the southern exposure of your house'. He ends his warnings against optimism lightly: 'For the present, why get excited about owning the

Firewatcher at work in New York city: how he was meant to see the glare of fire against the blaze of lights is not clear.

moon? What would you do with it, anyway?' Such high-profile articles, the designers' exposure at the World's Fair and their growing wartime experience, improved the professional recognition of designers, a fact acknowledged, unfortunately, when the New York Tax Service decided to make demands for tax under the unincorporated business rules in respect of the industrial designers' earnings. The designers argued that they were not selling goods but, like an architect or lawyer, providing services in a personal capacity. The threat of a half-million dollar tax bill made the designers close ranks, and Dorwin Teague was chosen to present a test case in 1943. His lawyers were able to argue successfully the parallels between industrial design and the other established professions, and cited Teague's recently published *Design This Day* as evidence of the seriousness of the profession. Loewy, who had been involved in preparing the case with Teague and Dreyfuss was reported to have made the horrified comment that thereby Teague (and not presumably, Loewy) became officially the first American industrial designer!

One part of the tax ruling stated that exemption was permissible as 'eighty percent of the gross income is derived from personal services actually rendered by the individual.' So whatever the different styles or approaches of the three main offices, in each case the leading designer was still perceived as the prime earner. Despite the increasing market for industrial design work – between 1935 and 1945 Loewy's client list, by his own account, grew from a dozen firms to over seventy-five – the businesses remained very much dominated by their founding personalities, whatever the actual workload below decks. The day of the acknowledged design team had yet to dawn.

The income tax case also led to the industrial designers considering whether to set up a professional body. There already existed an American Designers Institute, that had grown out of an association of designers working in the furniture industry and then extended its scope to include designers in other craft-based industries. For whatever reason, Teague, Loewy, Dreyfuss and others decided in August 1944 to found a Society of Industrial Designers, based in New York. Teague was president, Dreyfuss vice-president, and Loewy chairman of the executive committee. Harold van Doren was treasurer, Egmont Arens secretary. The rules defined candidates for membership as either those who were 'successfully engaged in industrial design' or academics teaching industrial design, while a designer was defined as someone who 'has successfully designed a diversity of products for machine and mass production'. Deliberately or not, this clause excluded many of the members of the ADI, who tended to work in single industries, and often within a craft tradi-

The Society of Industrial Designer's logotype, designed in 1946.

tion. Indeed, at the first meeting George Nelson and Charles Eames (the latter's plywood furniture had recently won an award from the Museum of Modern Art in New York) were not selected as members, because of these criteria. The rules were to lead to much friction between the two bodies in coming years, but what is relevant about the establishment of the society, and about the tax case, is the light they shed on how the designers at that time perceived themselves. The emphasis on personal skills, and an independent readiness to serve a diversity of clients, is particularly interesting. The contrast between the 'have design, will travel' approach and the corporatist approach of in-house design is a telling one. Much later, in 1959, Raymond Loewy was to speak at a meeting of the American Society of Industrial Designers against specialization in design: 'if designers get reabsorbed, ingested, digested mutated or re-orientated by...non-designing forces or executive enzymes, there will be no industrial design profession.' Even in 1979 he talks disparagingly of in-house designers. 'Captives! Following the orders of marketing managers makes them social servants.' In insisting on the independence of the designer Loewy was in agreement with his peers. Dorwin Teague's book implicitly treats the industrial designer as polymath, telling of 'this new profession of industrial design, in which one man of restless mind and many interests assembles round him a group of variously trained co-workers... and directs their efforts in an astonishingly wide range of activities.' He concludes: 'they – the industrial designers – are in fact a product of the times, and the fruits of a trend quite as definitely as they are leaders of that trend.'

At the end of the war America had hopes, and had money – one estimate suggests that the combined savings of Americans over the war years totalled an astonishing one hundred and fifty billion dollars or more – but there was little to spend it all on. Many manufacturers, as Loewy had predicted, were not able to retool immediately for old products, let alone new designs. At the same time federal efforts were made to keep industrial production at reasonable levels, to head off inflation. The war had caused immense social upheavals, not only because of the diversion of the workforce into the armed forces but also because of the war industries, when whole towns grew up around new industrial plants in a matter of months. The late 1940s were therefore a period of furious, but managed, activity, with housing and roads a top priority. Many designers and architects put forward their proposals for future housing – one such project, designed by Marcel Breuer, was built in the garden of the Museum of Modern Art, bringing the International Style down to earth in suburbia (the house was for an office worker who 'commutes to a so-called 'dormitory town'.') Buckminster Fuller, whose Dymaxion housing ideas, based on grain storage bins, had been taken over by the US Army as immediate and portable shelter systems, launched the Dymaxion house as a venture with the Beech Aircraft corporation. Henry Dreyfuss also worked on plans for prefabricated houses with Consolidated Vultee in 1947. Many of these initiatives – Fuller's among them – came to nothing. But if the idea of the prefabricated house failed to win popular support, the application of production line planning to the housing market did succeed. Entrepreneurs like Levitt and Kaiser used factory production methods to build low-cost houses rapidly. The resulting new suburbs may have looked monotonous and uninspired, but by such means there were well over one million housing starts in 1950 alone. And all these homes would need to be fitted out for heroes with fridges and cookers, tables and chairs, lamps and bathtaps. The prospect of plenty beckoned the designers.

TIME

THE WEEKLY NEWSMAGAZINE

DESIGNER RAYMOND LOEWY

He streamlines the sales curve.

striking it lucky

'The wonderful dreams that had kept me going throughout the hostilities had finally come true.'

Raymond Loewy *Never Leave Well Enough Alone*

In 1949 Raymond Loewy achieved a major first, in any businessman's terms. He got his picture on the cover of *Time* magazine. (He nearly scored a double four years later, when the Studebaker Starliner was also featured, but with the president of Studebaker, Charles Vance.) He is said to have joked to his publicity director, Elizabeth Reece, that 'all she had to do was get him on the cover of *Time*': it looks like she did achieve the impossible. That *Time* should have featured an industrial designer is understandable. There had been some disappointment in 1945 and 1946 that the expected flow of new post-war consumer goods had not occurred, but by 1949 the post-war boom was well under way, with nearly a million new houses being started each year, all needing stoves and fridges, furniture and light fittings. All the automobile factories were launching new models annually, from the Ford Fairlane to the Plymouth Business Coupe. The industrial designers were seen as being among the orchestrators of this plenty.

As to the choice of Raymond Loewy as the featured designer, there is little doubt that of the 'big three' Loewy was the most assiduous in courting the press, at any and every occasion: his 'genial knack for pushing himself to the front' probably served him well here.

The cover of *Time* magazine, October, 1949.

Above all, here was the designer with whose designs – for cars,trains, planes and household goods, the average *Time* reader would be most familiar: the subtitle to the cover is 'he streamlines the sales curve', and arranged behind Loewy's portrait are the SS Lurline, a cruise liner recently designed by the office, the SS-1, the Studebaker Commander Regal, the Greyhound Silversides coach, Frigidaire stove and refrigerator, the Hallicrafter Radio, Lucky Strike pack, the Eversharp fountain pen, Fairchild F 46 aircraft, and so on.

The article follows the familiar day-in-the-life pattern, with Loewy waking up to a Loewy-designed alarm clock, brushing his teeth with Pepsodent from a Loewy-designed pack, shaving with a Schick electric razor (designed by Loewy) and on to his office in the latest Studebaker, customized with a plastic tail fin, and a gold air scoop in the bonnet. Here the staff of 143 are mentioned, and an impressive list of current and forthcoming clients as well. Though Bill Snaith gets a mention, along with Baker Bernhart and the business manager Jim Breen (who had joined Loewy in the 1930s), the emphasis of the article is almost entirely on product design. The article describes his as the biggest design office in the USA, and the dominant one in the field: his fees, the $50,000 for Lucky Strike, for example, are listed, as is his projected turnover of 3 million dollars, and Loewy's own expected income of $200,000 or more. The Loewy lifestyle is also fully described: the houses in France, including a new apartment on the Quai d'Orsay in Paris, the house at Palm Springs, the lavishly – and originally – decorated Manhattan apartment. The work of other designers is also mentioned, and an estimate given of how much US manufacturers expected to spend on 'improving the way its products look': 500 million dollars. This is

Superstructure design for a McCormick line steamship, 1955.

clearly not only on design fees, so it is difficult to estimate what percentage of the market Loewy Associates 3 million represents, but in among the euphoria for design and for the success of 'suave, grey-haired, medium sized (5ft 10 in) Loewy' (his face is 'reposed, gentle, sad and as inscrutable as that of a Monte Carlo croupier') comes one important qualification: 'many big companies... have long since built up design departments of their own, but smaller companies, who cannot afford to do so, must depend exclusively on free-lance specialists like Loewy.' While Loewy's business continued to grow through the 1950s, with government departments added to private clients, the basic truth of that sentence could not be resisted: the day of the all-embracing design company, ready to tackle anything from a department store (*Time* noted that Loewy Associates had completed the first stage of refurbishment of

The Loewy house in Mexico.

Gimbel's in New York) to a tin of beans (Loewy Associates' label redesign of the Armor range of canned food gets a special mention), that day was coming to an end.

One of the headlines in the *Time* article describes Loewy as 'The shy salesman', drawing attention to his personal reserve – he was always Mr Loewy to his colleagues, and few seem to have known him intimately. In *Never Leave Well Enough Alone* he refers directly to his shyness with women, as well. In fact he was married twice, the first time in 1931 to Jean Thomson, who worked in his office. They had no children, and were divorced amicably in 1945: she remarried, but as Jean Bienfait she remained a partner in Loewy Associates until

Viola Loewy: this photograph was used on the dedication page of *Industrial Design*.

1950. In discussing the marriage in his biography, Loewy juxtaposes the mention of the failure of the marriage with an account of the wartime visit of two evacuee children from England, and the pleasure he got from this temporary and surrogate fatherhood. Whether the juxtaposition of the two stories was deliberate or not it is impossible to say.

In 1948 Loewy married Viola Ericson; they had one daughter. Contemporary photographs show Viola as a beautiful and vivacious woman, and she was to play an increasing part in Loewy's business. She was to define her own role as a business manager, dealing with the details of contracts and handling client contacts, relieving the pressure on her husband of detailed sales negotiations. Many years later, Loewy explained how he would 'go into training' before major meetings with important clients, and Viola's own testimony shows how important her supportive role was in preparing major presentations.

By the end of the 1940s the first generation of industrial designers had grown up, and the second generation, Charles Eames, Eliot Noyes and others, were hard on their heels. The Loewy office remained extremely successful throughout the fifties and into the sixties, nonetheless, but this was less due to the product design that had started and sustained the business during the thirties and forties. Architecture and in particular shop design had grown into a key part of the business. As early as 1934 Loewy had built a 'designer's office' for the Metropolitan Museum of Art Industrial Art exhibition, and some of the work on the *Princess Anne* ferryboat involved remodelling the interiors. In 1936 he was engaged on another marine project, together with a naval architect, George S. Sharp. This was to design interiors for three small luxury passenger steamers for the Panama Steamship Co., which sailed between New York, Haiti and the Panama Canal. The three, the *Panama*, *Cristobal* and the *Ancona* were fitted out in a contemporary style, with plain modern seating and discreet light fittings. The same calm – not to say bland – modernism is found in another Metropolitan Museum of Art interior set, the room for a five-year old child he exhibited in 1940.

More importantly for the architecture and interior design aspects of his business, in 1936 William T. Snaith had joined Raymond Loewy Associates. Snaith had studied architecture in New York and Paris, and had been working as a stage designer. His first job was to redesign the stationery department of a New York store: another early project was the design of

Today *Time*, tomorrow *Der Spiegel*: Loewy's book was an unexpected bestseller in Germany.

Lyons Corner House in London for the first London office. Snaith and the team in London were not only responsible for the architecture and interior design, but for fittings, graphics and uniforms – the nippy was born on a Loewy drawing-board. A 1949 photograph of Snaith shows a round, open face, the smile only partly hidden by a moustache. His pose is cheerful, even jaunty. He was to build the store planning and design department into the most profitable part of the New York office. Many of the designers who worked for him, including his closest assistant, Maury Kley, are still working in the same field.

Dorothy Shaver was president of Lord and Taylor, the New York department store chain.

This china service 'Undine' was one of several Loewy designed for the German firm Rosenthal.

Lord and Taylor's enlightened attitude to design has been seen in the 1934 *Industrial Design* survey, and Loewy had in 1937 redesigned Dorothy Shaver's fashion store. Then, in Loewy's words, 'we convinced a client that a store was an implement for merchandising and not just a building raised around a series of pushcarts.' The result was the commission to design a complete new store in Manahasset, Long Island. The choice of location is also important: Snaith had been arguing that department stores had to move to where their customers were, and affluent middle-class Americans were increasingly moving out of the cities into new suburbs. It is a pattern that is familiar today, and Loewy and Snaith were

This De Luxe Dispenser for Coca-Cola was an important post-War commission for Loewy.

among the first designers to appreciate its consequences for store planning and design. One consequence of the change to suburbia was that the customer went shopping by car: there was less need to show goods in windows: instead the shop name was placed prominently on an outside wall. Loewy instead aimed to make the shopping experience as pleasant as possible. In the Manahasset store, formal straight counters were replaced by 'Little Shops', placed informally, and with curved counters. 'The store itself becomes American Suburbia's village green.' Stock and fitting rooms, as the plan shows, were arranged outside this area, against the main walls of the building. Another new feature was the daylight selling area, a window

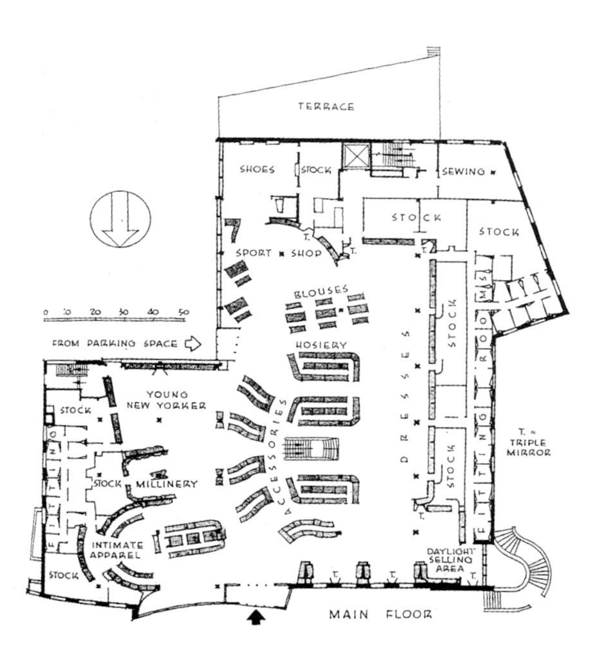

The floor plan for the Lord & Taylor's store, Manahasset *(left)* and the exteriors of two stores, at Milburn *(facing, above)* and Manahasset *(below)*.

looking into the sales area of the store, which the shopper passed on the way to the front entrance. Through the window customers could see other customers 'at work', rather in the way that newspaper photos of the crowds at the winter sales encouraged – supposedly – more bargain hunters to come along. The logic is interesting, though whether the window worked as planned is impossible to say. Certainly the design worked well enough. Lord and Taylor used Loewy and Snaith for a further six branches, the last of which opened in 1959.

During the war years Snaith had the opportunity to study in depth the future of store design, rather as Loewy and Virgil Exner had time for reflection on the future shape of the automobile. The Associated Merchandising Corporation commissioned a study from

Snaith, issued in 1944 under the title of *The Store of Tomorrow*. According to Loewy's *Never Leave Well Enough Alone*, Snaith pointed to four main conclusions. Firstly, the department store was 'probably the last of the manually operated large industries'. Secondly,

Foley's department store in Houston, Texas.

it was difficult, though not impossible, to mechanize the industry. Thirdly, high labour costs gave store operators little room for manoeuvre in meeting price pressure, Fourthly, the operation of most department stores was impeded rather than helped by the buildings in which they had, to date, been housed.

The chance soon came to put the blueprint for a new store into practice, with the commission from the Foley Brothers for a new store in Houston, Texas. Here the major effect of the new ideal was a store with no outside windows at all. The stock handling system, taking stock up from the basement to stockrooms behind the sales areas, and taking wrapped purchases down to the garage area, either for packing in delivery trucks or for collection by shoppers, could only work efficiently if it was arranged to run along the outside walls of the building, fed

P... ...teu aircraft interior for
L...ckheed.

by and feeding the central sales points. The architectural effect of the windowless exterior walls is indeed dramatic, but that was not the point. Mechanization would speed service to the customer, and reduce labour costs. The intended result was a happy customer and a happy client, a duality which sometimes seem to be the Loewy watchword. The same process of planning the store interior for greater efficiency and service with less manpower was applied to the redesign of the Gimbel's department store in New York. The professionalism of this approach by Snaith and his team is a far cry from Loewy's advice in the 1930s to Horace Saks of the Saks department store in New York to dress his elevator attendants in uniforms and white gloves, but there are some similarities, in Loewy's attempts to make the customers feel that shopping was a special event, and in giving the elevator boys and girls some esprit de corps.

Some years later, Snaith prepared another report, this time for the Supermarket Institute, on the future of supermarkets. Here market research, based on questioning large numbers of shoppers at different locations, suggested that the problem with supermarkets was not the easy availability of plentiful choice, but boredom with the repeated rows of displays. The solution proposed was to break up rows of shelving with specialized displays, presenting goods more theatrically. This goes back to the idea of shops within a shop that had been pioneered at Manahasset. Writing in 1979, Loewy listed over twenty major stores and chains as Loewy clients, proof of the success of this aspect of his business, though his praise for Snaith (who had died during open-heart surgery in 1974) is considerably more muted than it was in his earlier book *Never Leave Well Enough Alone*: but Snaith had been a driving force in the company equal to Loewy himself.

Other architectural projects included rail stations, a factory building for Fairchild (for whom Loewy designed the interiors of the F 47 airplane), and supermarkets for the Lucky chain and for Dilberts. The most prestigious project, perhaps, was the interior design, in 1952, of the Lever Building in New York, for which Skidmore, Owings and Merrill were the architects. Here individual colour schemes were devised for the subsidiary companies, which occupied individual floors in the building. These colour schemes in turn complemented the main scheme used in common areas such as the lobby: the American critic Lewis Mumford wrote, in the *New Yorker* of the interior 'I don't know any other building in the city in which so much colour has been used with such skill and charm over such a large

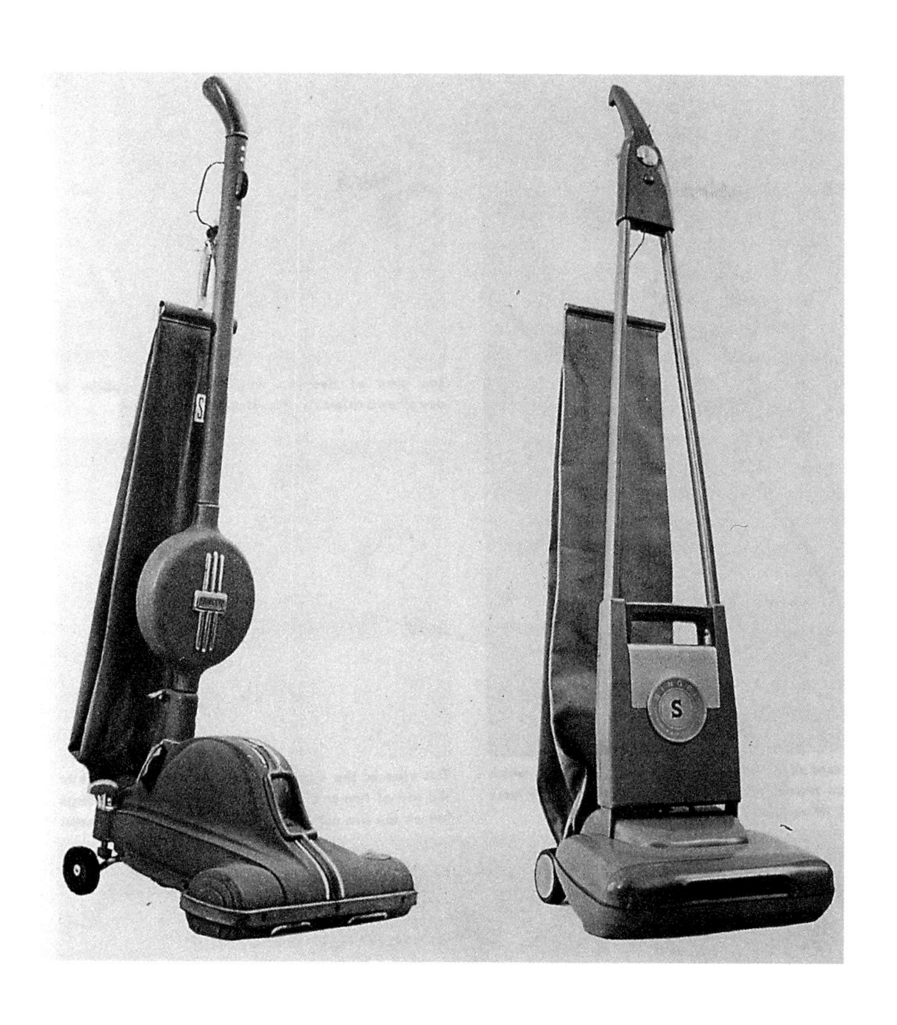

The Singer vacuum cleaner before and after redesign – 'Napoleon's hat pollinated by a horseshoe crab' was the description of the earlier design.

area.'

The other Loewy office in America that was a major profit centre was the Chicago office, headed by Franz Wagner. Here the packaging division was particularly important, as it was to

Loewy's work for International Harvester included a modular design for their service centres *(left and facing, above)*, as well as a new corporate logo *(facing, below)*.

be for the London and Paris offices. Loewy had been involved in corporate identity and packaging work at an early stage. His redesign of the Lucky Strike pack has already been mentioned, and another important pre-war account was International Harvester. For them Loewy began by redesigning the Farmall tractor, widening the front wheelbase to increase stability, improving the seating and controls, and improving visibility. Two years later the Caterpillar tractor was redesigned, again improving the seating, and otherwise cleaning up the outlines of what had been a purely functional design. At his own suggestion, Loewy also redesigned the company logo, doing so on the back of a dining-car menu in the train on his way back to New York. (Loewy once said that industrial design was ten percent inspiration, ninety percent transportation, and here the two seem to

have co-incided.) 'The spur of the moment creation of this trademark', he wrote, 'and its subsequent longevity contradict the notion of other designers that designing a new trademark always demands thorough, lengthy, expensive market research and a great many interviews,

tests and polls... These things may sometimes be necessary, but it all must start with an inspired, spontaneous idea.' Indeed in designing the packaging and labelling for the International Harvester range of spare parts and accessories – a list of over seventeen hundred items – the need for research and planning was evident, and the solution was based on a careful analysis. Loewy's office also designed a modular system for International Harvester's service and sales centres. The clean lines and clear layout of these buildings were very successful: over 1,800 of them were constructed. Together

The redesigned tractors for International Harvester: the Farmall *(facing)* and the Caterpillar *(above)*.

with the modern and abstract logo, they promoted the idea of International Harvester as an up to date, efficient company. Loewy's description of the services his company could offer a client, the mythical Widget Corporation, in *Never Leave Well Enough Alone*, includes not only remodelling their product, but also researching their market share, their competitors products, and looking at their advertising material and corporate literature. In this he is describing a complete design function, and not just the services of a stylist. His work for International Harvester was closest to this ideal, among his pre-war clients, though the range of work done for the Pennsylvania Rail Road runs it a close second.

The London office of Raymond Loewy Associates was in fact opened three times, firstly in 1934, then closed during the war, then in 1947, only to be promptly closed because of post-war restrictions, and finally in 1969. The third opening was initially to provide a European-based design consultancy service for Nabisco, the biscuit and cereals company, whose packaging the New York office had recently redesigned. Patrick Farrell, who had been working with Loewy on the NASA project, was asked to go to London to oversee reopening the office, and indeed at the time of writing still runs the Loewy Associates with his partner Thomas Riedel.

At the time Patrick Farrell left New York, in 1968, the Loewy offices employed over 200 people. The client list had grown from a dozen in 1935 to seventy five by 1945 and over one hundred and fifty by the end of the decade. This pattern of expansion, however, needs some qualification. It would seem that the emphasis, even in the glorious late forties, was shifting from product design, which had made the company's name in its early years, to store design and packaging. Secondly, the really big potential clients, whom Loewy had always personally identified as being the best targets for his services, had by the late forties already established in house design staff, and did not need outside help. Take the case of the motor industry: Ford, General Motors and Chrysler had long had their own design departments, fenced off from the rest of the company by walls of secrecy. Detroit culture was against letting their own staff see the designs, let alone outside designers and potential rivals. The fact is that these Chinese walls also cut the designers off from reality, and led them to evolve design ideas that were less and less associated with actual demand. It is not surprising that such a culture produced such a monumental bloop as the Ford Edsel: what is surprising is that it did not produce them by the dozen. Detroit culture, embodied by Harley Earl, styling supremo at General Motors until 1959, was elsewhere giving ground to a different discipline, in design-aware companies, of integrated design, in which the designer, still part of the company, was not isolated from it. In simplistic terms, this approach, characterized by Eliot Noyes' success at IBM, for example, is the victory of the European idea of 'serious' design, relating form to function to materials, as opposed to the slap-on-the-modelling-clay restyling favoured by the early industrial designers. More importantly, it was a design culture, not a design contact: the designer was to become an integral feature of the firm, involved with all processes of product development. The days of the designer winning the deal by catching the ear or eye of the president of the company were soon to be over. This sea change in design was, with hindsight, bound to come about as design became more and more accepted as a necessary part of the industrial process, and some design companies were to be better fitted to handle it than others: Dreyfuss Associates, for example, was long used to a research-based approach to design problems, as compared to Loewy's apparent preference for an instant solution, carefully studied and refined thereafter.

How then, did the Loewy office work? There are various accounts to chose from. Loewy's own is contained in chapter six of *Never Leave Well Enough Alone*: two earlier chapters have taken the reader along for a ringside seat at various design meetings (including getting stuck in a snowstorm at Chicago airport, and losing the chief designer to the charms of a singer in a nightclub). Another is in the 1949 *Time* article on Loewy, which paints a picture of him living in a completely Loewy-designed environment, from his toothpaste to his town car. What is clear is that individual projects were in the charge of divisional heads of offices, while Loewy himself got down to selling the company's services to clients. But Loewy himself kept a careful eye on everything that went out of the office. Every drawing, whoever made it, that was to be shown to a client, was seen and signed by him, for example, leading Jay Doblin to comment that the best drawings thereby always ended up in the waste paper bin. While there was never a Loewy style in the sense of a particular look all Loewy-designed objects would have, there was a Loewy approach to design, and his colleagues were expected to learn from Loewy, so that the public face the office presented to the outside world would be a cohesive one. What drew designers to work for a company so dominated by the spirit of its founder (who always retained a majority shareholding in the company) was, according to Elizabeth Reece, the opportunity to work for the largest company in the field, on a wide range of different projects. Several notable designers worked at one time or another for Loewy: Gordon Bunshaft, later a partner at the architects Skidmore Owings and Merrill was one, and Jay Doblin, later founder of Jay Doblin Associates in Chicago, was another, to name but two. Bill Snaith's launching of various store designers on independent careers has already been mentioned. Loewy himself once claimed to run the best design school in the USA!

What is abundantly clear is that Loewy brought a great flair for presentation and promotion to all he did. His zest and charm still bounce off the pages of books and magazines, and these qualities must have been an enor-

Loewy's own cars were often redesigned by him: he was particularly proud of the Lancia Loraymo *(above)*, and of his Jaguar XK 120 with bodywork by Boano of Turin *(below)*.

mous asset in the busy, boosting America of the 1940s and 1950s. In photos from the time he appears dapper but dynamic, the half inch of white shirt-cuff always visible (*Time* was careful to point out how hard the designing life was on a designer's shirt-cuffs, explaining that Loewy had his shirts made to measure to cope with this.) He has a purposeful look, but not

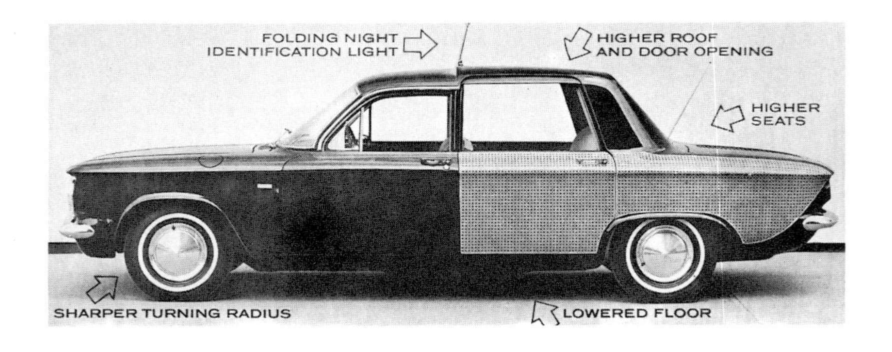

perhaps that of a regular good fellow as Babbitt would have said, for there is a foreign air about him. Perhaps it is the moustache, or something in the set of the eyes. The combination certainly seems, for a time, to have worked. According to Elizabeth Reece in the recent *Loewy* catalogue, Bill Snaith once described Loewy as having 'unerring vulgar taste', by which he seems to have meant an ability to recognize what the public would want once they saw it. The success of his post-war Studebakers supports this, and his 1930s locomotive designs seem to sum up the essence of the streamlined age.

But the taste Loewy believed himself to have was a different one, anything but vulgar, even in the nicest, Snaithian sense of the word. Nowhere is this better shown than in his series of

Other redesigns over the years included the Lincoln Continental town car *(above, top)*, a Ford Corvair project *(above, left)* and rebodies for a BMW *(facing, top)* and Cadillac *(facing, below)*, the latter two by Pichon Parat in France.

private cars. He seems to have had almost a mania for taking other manufacturers' cars apart and redesigning them, either as town cars or sports cars. The list comprises two Lincoln Continentals, two Cadillacs, one Studebaker, one BMW, one Jaguar and one Lancia, at least. This is apart from the Hupmobile and the Studebaker Starliner that he redesigned at his own

Raymond Loewy explaining the Loraymo and its airfoil to General de Gaulle.

expense to show his ideas to the respective managements. The Lancia Loraymo (the name is based on his international telegraph code, Lo-ewy Raymo-nd: he also used it to name several of his yachts) earned some scathing comment – particularly for the adjustable airfoil on the roof, according to Bruno Sacco of Mercedes, writing in the recent *Loewy* exhibition catalogue. Loewy's claim, at the Paris Motor show where the Loraymo was exhibited, that the airfoil was to prevent the car becoming airborne, and that the idea was later taken up by major racing car designers, is to say the least overstated: the best comment on it is the illustration of Loewy 'explaining the airfoil to De Gaulle': mon général looks distinctly unimpressed. In most cases the line of the car is improved, along the lines of less weight and less chrome favoured by Loewy. Taken as a whole, the series of car designs are impressive, but they have also a boyish quality about them: they look too much like illustrations, and too little like real cars. And they overlook the fact that automobile design, both before and after the gas-guzzler days, is a total process, not just a matter of styling. In these cars Loewy's relentless individualism and showmanship coincide with his vanity. When lecturing, he would emphasize his criticisms of Detroit styling by showing details of tailfins and radiator grilles photographed together with a small white French poodle. This undercutting of the body armour of the American dream is surrealistically funny, but misses his point. Much of the automobile design he himself did, whether for his customized private cars or in his professional designs, was simply putting a new shape onto an old chassis, a shape motivated by 'European' ideas rather than Detroit ones, but it was still only styling.

the final frontier

'Some post-war theorists are even promising us that old chestnut, the interplanetary rocket ship'

Raymond Loewy, *New York Times*, September 1943

The front cover of *Industrial Design* shows Raymond Loewy inside the mockup of the Skylab space station. At seventy-five, his hair has gone white and his face is somewhat lined, but is moustache is as dark and trim as even, the collar as wide and white, and the suit as smart. The space projects on which Loewy worked for NASA were the final achievements of his career, and were, from his speeches and writings, immensely important to him. The cover photograph is carefully posed and framed: the setting is the mess/galley of the spacecraft, and behind him are the food cabinets, to his left the porthole, and in front of it the triangular galley table. These three features summarized in physical form his contribution to the Skylab project, and the choice of this particular image as a cover picture is surely very deliberate.

If the project was so important to Loewy, it is surprising to discover how many inaccuracies and half-truths are to be found in the text he himself wrote about his involvement. To take the most glaring example, Loewy implies that he had been in touch with Jim Webb of NASA since being introduced to him by President Kennedy in 1962, and that 'the call from NASA' resulted from this contact. In fact, though NASA was involved in the decision to invite an industrial designer to advise on Sky-

Raymond Loewy wearing a spacesuit on a visit to NASA.

lab, the invitation to Loewy came from the Martin Marietta Corporation. From Reyer Kras's excellently researched article in the *Loewy* catalogue, the complete narrative emerges. NASA was almost indifferent to the design aspects of the project: it was only when George Mueller, director of Office of Manned Spaceflight, saw the mockup of the original 'Wet' Skylab interior that the question of improving its habitability arose. (The initial Skylab project was to build the Lab in a fuel tank in the final stage of the rocket, and convert this into the lab once the fuel had been used up. Later the 'Dry' idea developed, which was the one finally used. This also involved making the fuel tank section of a rocket the framework for the Lab, but not using it to carry fuel on the way up.) Loewy's first report to the Martin Marietta Corporation dealt with the design failings of the first Wet project: the confined spaces and projecting surfaces, the caged effect of the aluminium grilles used as separators, in an effort to save weight. The report has rather too much to say about Loewy's personal expertise as a transport designer, and ludicrously refers to his Rocketport as confirming his early commitment to spaceflight! (Loewy had conveniently forgotten that in his wartime *New York Times* article discounting post-war design expectations he had spoken disparagingly of 'that old chestnut, the interplanetary rocket ship.')

With the adoption of the Dry model the Loewy contract was reconfirmed, and despite the limited nature of the brief Loewy began to shower NASA with suggestions. At the time little was known, either in medical or psychological terms, about the effects of prolonged orbital flight, in which astronauts would be required to work for long periods in zero gravity. Loewy claims several major successes for his team: an informal working uniform, flush

Loewy in the mess area of the Skylab mock-up: the cover picture for *Industrial design.*

surfaces for storage cabinets, a triangular – and so non-hierarchical – table for the mess area were among the minor points. His team also worked on the organisation of the workspace in the lab, and on colour schemes for the different areas of the lab: these last were certainly adopted. One of Loewy's main allies in NASA was Caldwell Johnson, the head of research on the habitability experiment: Dr George Mueller also seems to have supported his involvement. The two major proposals Loewy made could be described as looking at potential psychological problems. Firstly he became insistent that a porthole, allowing a view of Earth at times, was a necessity: 'above every other consideration', he wrote in one report, 'a porthole should be provided in the wardroom.' He also made a strong case for the provision of a

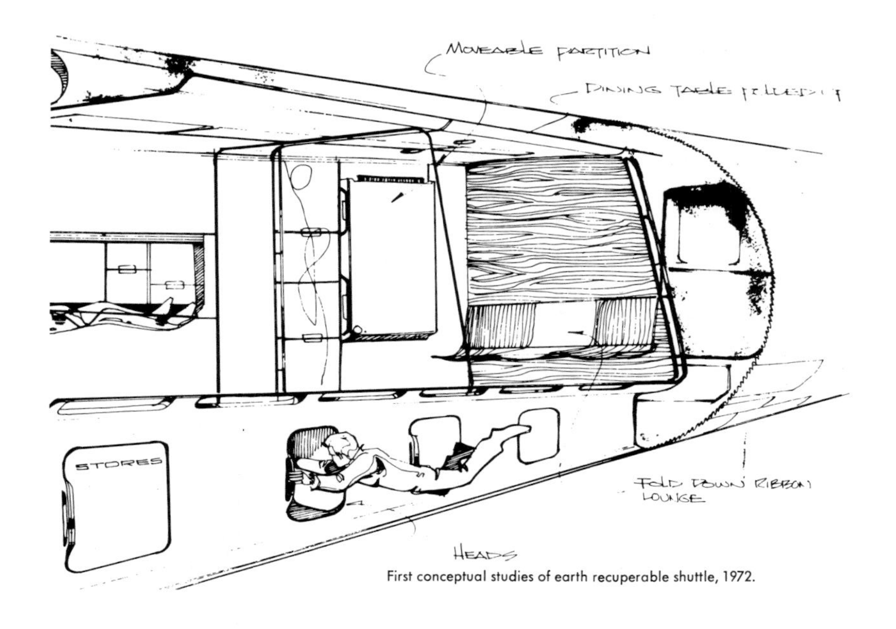

First conceptual studies of earth recuperable shuttle, 1972.

private area for each crew member, in which the crewman could 'sleep, relax, read or think in absolute privacy'. These two points also carried the day. The provision of any additional aperture in the outside structure posed considerable engineering problems. Loewy's pride in achieving these results is evident in his choice of the cover photo of *Industrial Design*, but, as Reyer Kras says, the interior environment was not mentioned either favorably or otherwise by the astronauts on their debriefing. Loewy claims in his book that 'all the Skylab crews... stated that without the porthole the mission might have been aborted.'

Tom Wolfe's *The Right Stuff*, with its account of the development of the Mercury and Gemini space missions from the original high altitude and high speed rocket-plane research of the late 1940s, gives a different viewpoint on

A drawing of crew's accomodation in the Orbiter space station *(above)* and *(facing)* an interior model for Skylab.

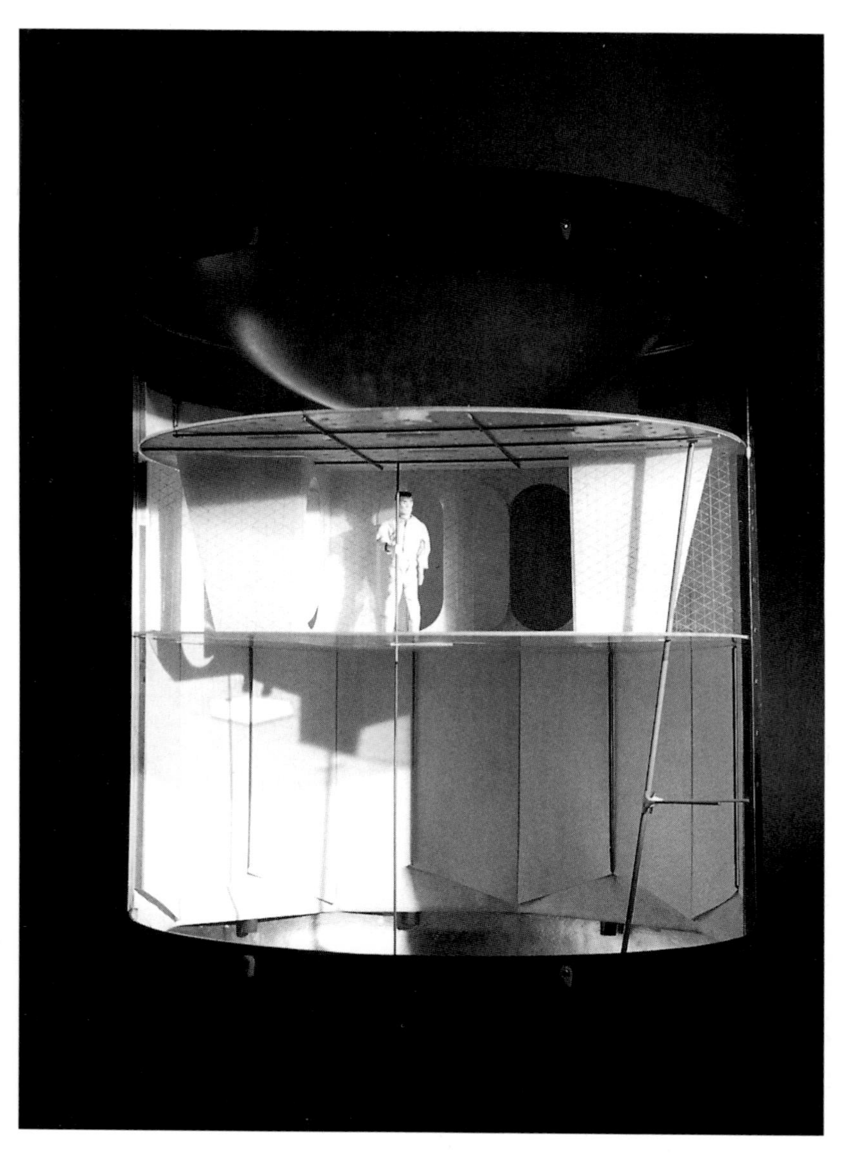

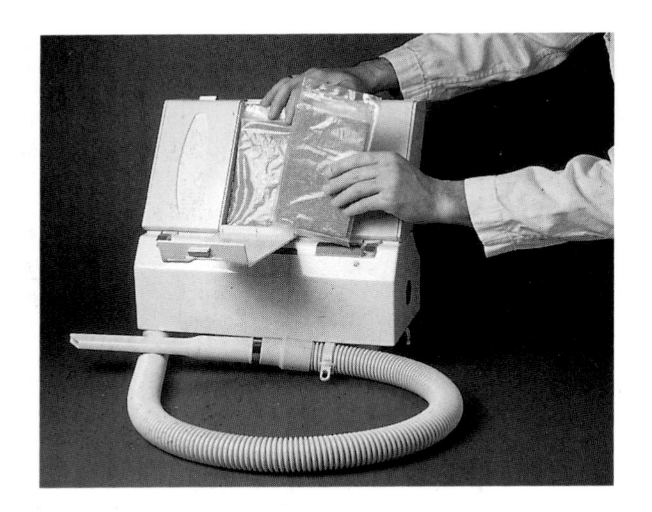

Equipment designed for space station included this portable cleaning unit *(above)* and a food tray *(below)*. Loewy is seen *(facing)* with a model of Skylab.

the relations between NASA and its astronauts. The NASA viewpoint, as it emerges from Wolfe's flashing narrative, was that the astronauts were an unfortunate necessity, to be treated and tested as an any other mechanical component. This viewpoint may be exaggerated in the interests of Wolfe's argument, which is to show how the astronauts gave the project a national and patriotic dimension,

with themselves centre stage as the leading warriors of the Cold War. However that may be, it is true that the astronauts themselves were used to performing difficult tasks under stress in complex surroundings, and their response to any amelioration of the space craft was likely to be practical rather than aesthetic. Unlike Captain Loewy and his well-pillowed dugout, the military test pilots recruited as astronauts took a stern view of their task. It is true that they had insisted on the provision of a window in the Mercury spacecraft, but this was a key part of their arguments that as pilots they were flying the craft: they were not just there for the ride. The question of a window in the Mercury spacecraft is highlighted by Tom Wolfe as one

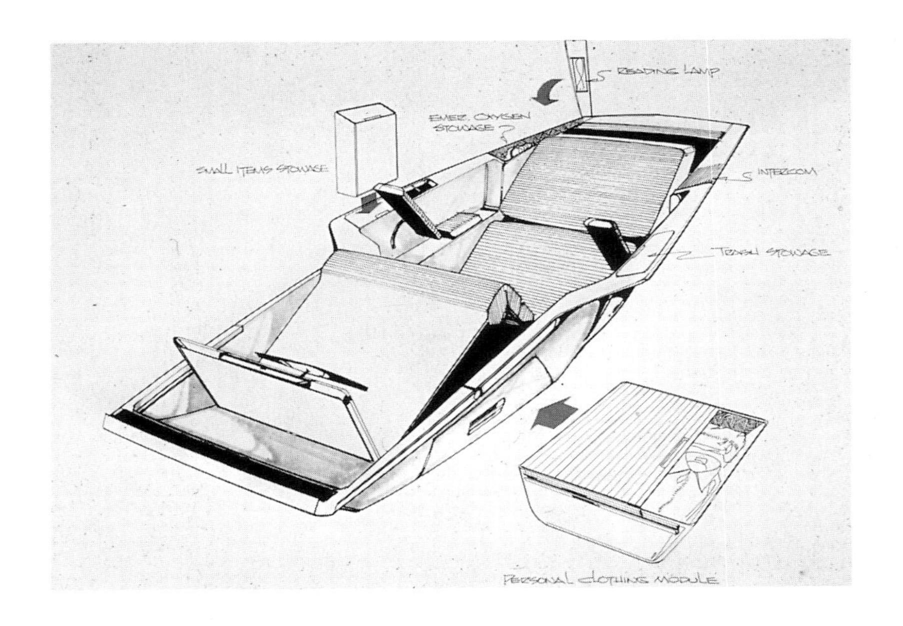

Astronaut's couch/chair proposed for Skylab.

of the major turning points in the story: so NASA was used to being argued down over portholes long before Loewy arrived on the scene.

From his work on Skylab Loewy went on to make suggestions for habitability on the space shuttle project, for earth orbit space stations, and for a space base project. In his published account illustrations from the different projects appear together, and it is difficult to separate the actual from the fanciful. A further photograph of Loewy in the wardroom of Skylab appears on the same spread as a suiting-up system for the space station, models for the interior of a multi-level space station, and a fanciful rendering of a 'space taxi'. Much space is also given to the 'fecal collection unit', though whether the designs for this essential function were for Skylab, the space shuttle or the space station is not made clear. The draw-

The Skylab porthole in use, from a contemporary television picture.

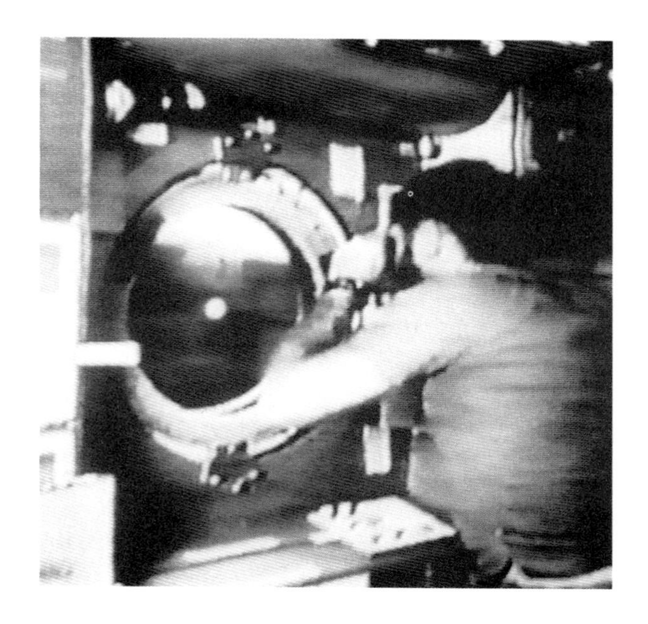

ings for the longer-term projects, the Space Station and the Earth Orbiter or Shuttle, seem to draw on science fiction rather than space fact. John Hoesli's sets for Stanley Kubrick's film *2001*, released at about the same time, have a considerably more convincing look to them. In one rendering, for example, although the fact of zero gravity is acknowledged in the placing of room openings at unexpected angles, the astronaut's stateroom is fetchingly panelled in wood veneer! According to Patrick Farrell, the designer responsible for the illustrations was a product designer rather than an illustrator, and the unfortunately amateur feel of some of the drawings may be due to this inexperience. Raymond Loewy tried his own hand at illustrating aspects of NASA's work, with a portfolio of prints that were later exhibited at the Jack Gallery in New York: the views of the moon's surface and of space vehicles in flight

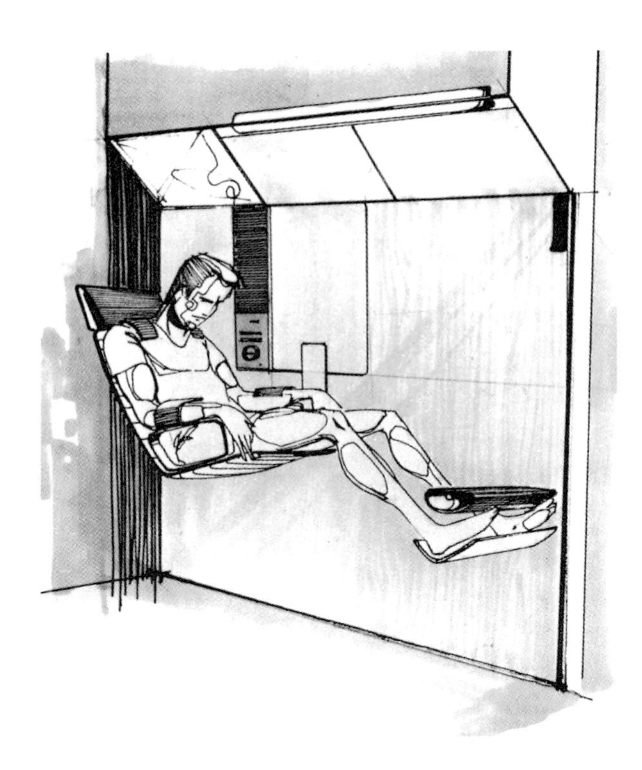

Two further views of life on the space station: a crewman in his private relaxation area *(left)* and the crew accomodation generally *(facing)*.

underline how important and precious this project was to him.

There is another image that sums up the almost boyish enthusiasm which infects some of the office drawings for 'extra vehicular activity' or nuclear-powered spacecraft. It shows Loewy in a space suit, surrounded by NASA officials, and with an immense grin on his face. The story, as told to Reyer Kras, is that once on his way to Cape Canaveral for a meeting on Skylab, he had his Lincoln – air-conditioned and chauffeur driven – tailed by his Studebaker Avanti. Stopping just before the Cape, he changed cars, and changed into the spacesuit, and drove himself to the base in the Avanti, claiming on his arrival that he felt he had to put

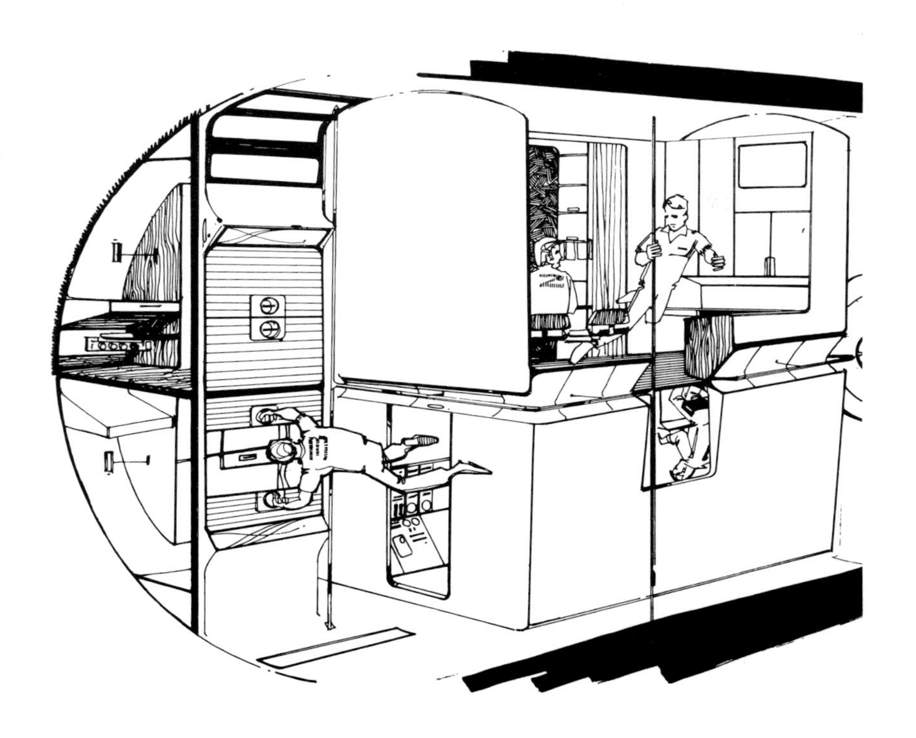

the suit to the test himself, by driving across the desert in it: after all, if the designer didn't do his best to understand the needs of our boys the astronauts....

NASA's rather cooler acknowledgement of the Loewy office contribution came in a letter from Caldwell Johnson. He praises the wide range of mock-ups, drawings and reports provided and the promptness with which these were produced. In particular, the Skylab mock-ups served to 'highlight specific aspects of habitability which are best understood in three dimensions'. Johnson ends his report by acknowledging the 'sound, professional support' received from the company. In his detailed list of nine projects, only three appear to have been

The overalls designed for use on Skylab – one project that was implemented.

definitely accepted: the coveralls for Skylab crews to wear, the accompanying Garment Storage Modules, and the colour schemes for Skylab. The designs for the Waste Management compartment were being tested full-scale in zero gravity flights, and various other proposals, including the galley on Skylab and the wardroom and staterooms on the Space Station, were 'under consideration': there is no mention of the famous porthole. Beneath the NASA-ese (Loewy called it 'orbit-talk') a simple fact is concealed. Before Skylab no-one, at NASA or anywhere else, had a clear idea of how human beings would stand up to prolonged periods of time in space. There was nowhere a start line for design thinking, beyond the engineering necessities of creating an atmosphere in a vacuum. For Loewy and his team it was a leap in the dark, and if their attempts to envisage all possibilities now seem odd or even trite, it should be remembered that they were starting from zero.

Loewy's version of this letter, in *Industrial Design*, makes two significant changes to it. The detailed project-by-project analysis is omitted, and so the impression given is that all the mock-ups presented had the same impact. Secondly, the name of the company, Loewy/Snaith Inc., becomes the Loewy Company in his version. Raymond Loewy's company had gone through a number of name changes: first incorporated as Raymond Loewy Associates in 1944, then as the Raymond Loewy Corporation in 1949. In 1961 the name became Raymond Loewy/William Snaith Inc. There are several possible explanations for this last change. Loewy was by then in his late sixties, and increasingly occupied with his agency in Paris. The contribution to the success of the business made by William Snaith's store design department was undoubted, and should be acknowledged: Snaith had traded on his success in building up the store design department to increase his shareholding in the company, so a name change acknowledged his success, and his ambitions, which were every bit as fervent as Loewy's own. The proportion of product and packaging design to store design changed rapidly in the 1960s, in favour of store design. With Bill Snaith's unexpected and early death in 1974 the business dwindled away. Loewy came back from Europe to take over the reins, and change the name back again. He also de-

A limited edition print by Raymond Loewy depicting a moonscape.

cided to put a banker, Jim Sheridan, in as president. This solution was not the right one, and the New York office was closed a couple of years later. Loewy let the client list go – mainly to Sheridan Associates, a company Jim Sheridan formed later, but Loewy retained the rights to the name. These eventually passed to the London office, which maintains an international presence, and the Loewy name, to this day.

The change of name to Loewy/Snaith was published on January 5th 1961, the day the New York Times headlined the breaking of diplomatic links with Castro's Cuba. The announcement claimed that the company's change 'reflects the need of its clients for integrated marketing and research counselling in addition to the company's industrial design services', and went on to point out that Bill Snaith has been the managing partner since

1956. Since Loewy in 1951 had been describing just such marketing and research services as part of the industrial design package offered to any customer – such as the fictive National Widget Ltd in *Never Leave Well Enough Alone*, this explanation poses more questions than it answers. Another story is that Loewy's habit of insisting that all drawings that went out from the firm bore his initials or signature finally became too much for his colleagues. In the 1970s something similar occurred, when Loewy tried to arrange an exhibition of 'his' drawings and renderings in a New York gallery. The gallery show was cancelled, with the designers who had actually produced the drawings complaining about the infringement of their moral copyright. This event shows Loewy overstepping the mark once again, as, one would now say, he had with his spacesuit stunt.

most advanced yet acceptable

'Industrial design... It's a simple exercise – a little logic, a little taste and the will to co-operate.'

Raymond Loewy, 1979

There was a different school of criticism growing in the USA in the years before and after the Second World War, which targeted not only Loewy but the ethos of all the first generation of industrial designers. This was a view of 'good design' that borrowed heavily from European ideas, and was in the 1930s and after the War much influenced by the Bauhaus émigrés to the USA. It was a view that found its focus in the Museum of Modern Art in New York. The Museum's annual exhibitions of good design, and its design collection, set a trend for a Modernist design ethic, essentially different from the market ethic of Loewy and others. The same colonial complex that Tom Wolfe ridicules in architecture, in his *From Bauhaus to Our House*, seems to have played a part in the definition of 'good design'.

MoMA's first director, Alfred Barr, had set out from the Museum's foundation in 1929 to mark its brief as widely as possible. The Modern Architecture exhibition in 1932, which coined the phrase International Style, the Machine Art show of 1934, and the Bauhaus exhibition in 1939 deliberately took the mu-

Raymond Loewy in a 1970 photograph.

The IBM Selectric typewriter *(left)* and its designer, Eliot Noyes *(facing)*.

seum into areas in parallel with contemporary art, and in so doing Barr grouped around the Museum some of the keenest thinkers and writers on art, architecture and design of the time, from Henry Russell Hitchcock and Philip Johnson to Eliot Noyes and Edgar J. Kaufman Jnr. Barr's own introductions to his exhibitions, such as *Cubism and Abstract Art*, 1936 and *Picasso, Fifty Years of His Art*, 1946 remain key statements of the Modernist position.

The industrial design exhibitions at the Museum of Modern Art fall into two groups, the pre-war exhibitions selected by the Museum's staff and the post-war series, rather more formally organized with outside selectors. The first exhibition of 'Useful Objects under Five Dollars' was put on in 1938, directed by John McAndrew, curator of architecture and industrial design. The exhibitions, often staged during the pre-Christmas period, were well received and were popular with the public as a source of ideas for acceptable and inexpensive

gifts. Though the intention was to foster interest in mass-produced – or at least series-produced – goods, price was no so important a consideration as quality: McAndrew, introducing the 1940 exhibition, stressed the aesthetic aspect, referring to the classic triumvirate of fitness to purpose, fitness to technology and fitness to material. In the same year Eliot Noyes arrived from five years under Gropius's tutelage at Harvard to become first curator of industrial design. During his tenure he organ-

ized an exhibition of 'Useful Objects in Wartime' as well as the important exhibition of 'Organic Design in Home Furnishing', in which Charles Eames and Eero Saarinen's furniture designs won a major prize, hailed by *Architectural Forum* as 'the most convincing glimpse of the future'. When Noyes returned briefly to the Museum after war service in Washington, he began plans for continuing the Useful Objects exhibitions, defining design quality as combining 'fitness for intended use, an intelligent use of materials, reasonable adaptation to the manufacturing process, and a contemporary aesthetic solution'. His successor, Edgar Kaufman Jnr., maintained the insistence on Functionalism and truth to materials by including battery jars and laboratory dishes in his first exhibition. Similar 'industrial' objects had been included in the 1934 Machine Art exhibition.

The Bauhaus tradition, that Walter Gropius and Lazlo Moholy-Nagy had transported to

Harvard and Chicago respectively before and over the war years, was, in part, the source for these assertions of the importance of rational simplicity and the avoidance of decoration. Mies van der Rohe's architecture was equally important, and his and Le Corbusier's design work equally cited, in developing the Modernist aesthetic in design. In a volume celebrating the first twenty-five years of the Museum, published in 1954, the cross-referencing to this European model is continuous: 'these objects depend for their appeal on clarity of forms and relationships, rather than on applied ornament': 'Machine Art... objects represent an attitude towards design so basic and powerful that even the twentieth century handicraftsman has come under its spell' and so forth, even claiming that the Bauhaus was a household word in prewar Germany! Other voices were making the same claims. Paul Rand, the graphic designer who worked with Eliot Noyes on the Westinghouse and IBM corporate logos, and

Ray Eames and Edgar J. Kauffman working on a Good Design exhibition *(above)* and a Good Design showcase display *(facing)*.

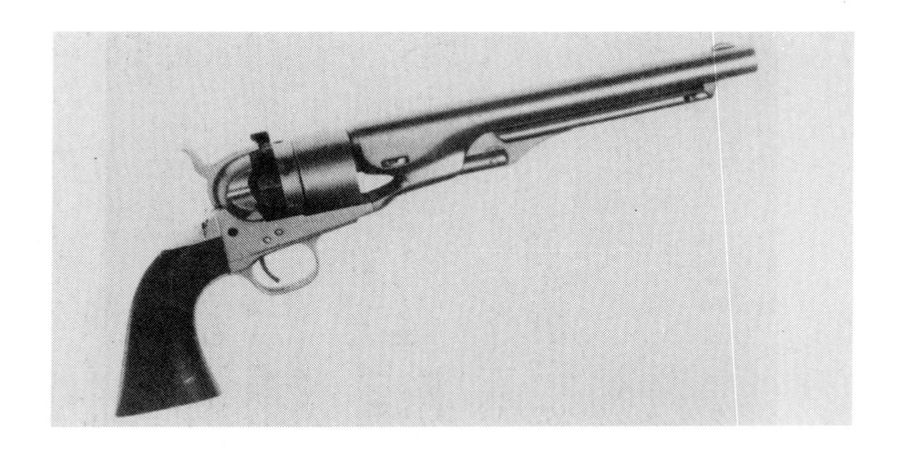

who in 1949 designed the cover for the Museum of Modern Art booklet *Modern Art in Your Life*, published his *Thoughts on Design* in 1947. He insists on virtually the first page that 'visual communication... should be seen as the embodiment of form and function; the integration of the beautiful and the useful', and he cites or illustrates the painter Mondrian, Moholy Nagy (who had been involved in restarting the Bauhaus in Chicago in 1938), the Swiss typographer Jan Tschichold and the English art critic Roger Fry – as well as the I-Ching – in the course of his short book. Eliot Noyes was also writing regularly in the 1940s in *Consumer Notes*, in a series of essays entitled 'The Shape of Things'. He regularly criticized adopting design styles from one era to another and poaching from one medium for another: refrigerator doors that looked like automobile bumpers, for example, or toasters imitating table silver – 'social climbers', as he called them. He was vigorous about American car design, calling it 'inept and trashily decorated', and criticizing the superficiality of the styling approach.

The good design school can point to a number of major American successes, that have be-

The classic anonymous design of the classic American object: the Colt Peacemaker revolver.

The Coke bottle is another American classic. The Loewy office did some work reshaping the larger sized bottle used in vending machines: this led to some ambiguities later, which are unravelled in Stephen Bayley's article in the *Loewy* catalogue.

come design classics – Charles and Ray Eames's furniture, for example, or Eliot Noyes's own Selectric typewriter for IBM. The 'Good Design' argument could have taken its title from Edgar Kaufman's series of exhibitions organized jointly with the Merchandise Mart, to promote good design, specifically by encouraging manufacturers to add modern products to their ranges. The selectors included, over the years, Eero Saarinen, D.J. Du Pree (president of Herman Miller furniture) and Serge Chermayeff, all of whom could produce proper Modernist credentials if required. Between this new generation and the design pioneers there were considerable differences of approach, which were in the end irreconcilable. At the beginning, as Ada Louise Huxtable has pointed out, both the 'popular designers and high art proponents had in common the romantic ideal of perfect products for a perfectible society.' With the end of post-war enthusiasms for a new era, and the acceleration of the American consumer boom, even this Utopia faded away, and the stylists and the men of principle drew up on opposite sides of the divide. Some products could still cross over: Eliot Noyes was very complimentary about Loewy's 1947 Studebaker Champion, for its originality of form, and the Museum of Modern Art added the Hallicrafter Radio to its collection, for example.

Against this background of good design versus market design, there is a third idea, the idea of true American design as essentially anonymous, consisting of forms in some way uniquely American – Colt revolvers, Coke bottles and Levi jeans have been cited, by Stephen Bayley, for example in the recent *Loewy* catalogue. The accolade given to unsung designs, such as the Douglas DC3 aircraft in an ICSID poll, is another example of this spirit, and a recent compilation of classic American designs includes the Fender guitar and the supermarket trolley, along with the Zippo lighter and Tupperware. This argument sometimes sidesteps the fact that such objects were in fact designed, even if the designer's name may be now unknown, and competed in the market with other designs (Remington's pistols and Pespi, for example), and the anonymity of their designers was not a cause of their success, or failure. But this approach – the object naturally evolved through a specifically American context and need – does have a comforting link historically with that quintessentially American event, the Shaker style. And in insisting on the Americanness of American design it draws parallels with jazz and American music, and with the ultimate architectural symbol of America, the skyscraper, by putting America first.

Between these two approaches, the Scylla of Modernism and the Charybdis of expediency, the streamlined look of the 1930s sailed safely, if not as dexterously as Ulysses. Whether called streamlining or cleanlining, the style by its nature appeared to eschew unnecessary decoration, it could be related to scientific principles of aerodynamics, it did provide a coherent and organic approach to the forms of new kinds of machine-made product – the car, the radio set, the telephone. It also provided a repertoire of ideal forms adaptable to different settings, from racing car to a building to a pencil sharpener (a projected design by Loewy described by Dreyfuss as an absurdity, and criticized by Philip Johnson and the Functionalists). As well as being undeniably American in its applications, the streamlined style could satisfy both the purists and the pragmatists, some of the time: Moholy Nagy, for example, perceptively realized the appropriateness of streamlined forms to some manufacturing processes, especially injection moulding. But the fact is that the motivations of the style were not necessarily in tune with either theory, any of the

THE SATURDAY EVENING POST

Dollar for Dollar

you can't beat the new PONTIAC

Everything it Takes to Make You Happy!

Only Car in the World with Silver Streak Styling

•

America's Lowest-Priced Straight Eight

•

Lowest-Priced Car with GM Hydra-Matic Drive

•

Thrilling, Power-Packed Performance —Choice of Six or Eight

•

World Renowned Road Record for Economy and Long Life

Everybody who has seen the new Silver Streak Pontiac is impressed by the *dollar for dollar* value this wonderful car offers.

Most people came prepared to see a beautiful car—for Pontiac is always an outstanding beauty. Pontiac's superb performance and dependability are well known virtues, too.

But the remarkably low price of the wonderful new Pontiac is big news. For here is America's lowest-priced straight eight. Here is the lowest-priced car with Hydra-Matic Drive—now even lower priced than ever!

If you haven't seen the new Pontiac yet, you should do so soon—you'll find it has *everything* to make you happy, at a price any new car buyer can easily afford. Dollar for dollar, you just can't beat a new Pontiac!

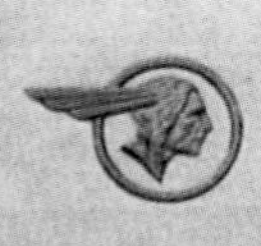

PONTIAC MOTOR DIVISION OF GENERAL MOTORS CORPORATION

The contrasting styles of General Motors and their European competitor are summed up in their advertising, for Pontiac *(facing)* and for Volkswagen *(above)* in a classic campaign by Doyle, Dean and Berbach.

time. The American design style of the 1950s, however, the 'borax and chrome' school, was completely beyond the pale. How does form follow function round a Cadillac tailfin, or design imperatives explain a vacuum cleaner with a light on it (does it *need* to see where it is going?). The webs of common terminology – organic forms, new technologies, adaption to social change – that had stretched to link the streamliners with the good designers in the 1930s broke up under the stress of coping with plenty in the 1950s: no sleight of logic could assimilate a car that looted its imagery, via Freud, from fighter aircraft and killer fish. In fact the American style of the 1950s was created by several of the same people who had created the streamlined style, and was motivated by the same kind of pressures, interests and ideas that had been at work twenty years before. The rush to produce a new model each year was not something that started with the 1950s, but had been endemic in American manufacturing since Henry Ford tooled up for the Model A. However, in the 1950s and 1960s, unlike in the 1930s, the whys and wherefores of design decisions were being scrutinized more and more, not only by academics and committed museum curators, but by an increasingly design-aware public. The failing of 1950s design was its absence of theory: while the designers went on with the job of making products look the way people wanted to buy them, the arguments had moved onto a quite different plane, the thinkers leaving the doers behind them. A similar split followed in the products themselves: one part of the market went for European-influenced – or European-imported – design, while traditional American design dominated the rest. The products were different, and the way they were promoted and sold was different also. A good example is in 1960s car advertising, where the classic Volkswagen campaigns by Doyle Dean and Berbach are in total graphic, semantic and conceptual opposition to the baroque approach of, say, General Motors. For the styling guys of Detroit slogans were better than theories.

Loewy's own design theory he called MAYA, for 'most advanced yet acceptable'. The punctuation and the grammatical usage of 'yet' in this phrase are interesting. Does the phrase carry a comma after advanced, meaning the design proceeds towards the most advanced possible but stops at the threshold of acceptability, or is it free-form, meaning that the level of advancement pushes the envelope of acceptability even further outwards? 'Yet' sets a limit in the phrase, but where is it set? A cynic might turn the phrase round, without losing much of its meaning, into 'most acceptable yet advanced', suggesting a slavish willingness to follow market forces, for example. And the other open question is whose level of acceptability is being invoked: the client's, the consumer's or the designer's? All MAYA shows what a handful a handy acronym can be. (The invocation of pre-Colombian culture also gives a spurious authenticity to the term.) Loewy's own analysis of the term is best described as rambling, and indeed he clearly enjoyed its chameleon character: if a client wanted a design philosophy, there it was to hand, in the colour or shape desired. He sets out, in *Never Leave Well Enough Alone*, a set of impressive-sounding principles, discussing product norms (as established by, well, established manufacturers), design gaps ('3. The risk increases as the square of the design gap between norm and advanced model in the case of a large manufacturer...') and MAYA coefficients ('Two unmarried individuals each having a high MAYA coefficient have a lower common coefficient as soon as the marry.') Much of this is clearly tongue in cheek, as when he writes 'd. The wife is often the deciding factor at the time of purchase. Her influence seems to decrease in direct proportion to the length of the marriage, reaches a plateau and then reverses itself in later years.' This meaningless wisdom and cod mathematics might have appealed to Loewy's executive audience, but do not take an understanding of his design much further, entertaining as it all is.

One way of analyzing the impact of a designer's philosophy on his work is to look at the progressive stages of a given design to see the impact of principles on ideas. For example, in Herbert Bayer's experimental type designs the Bauhaus principles of true form and legibility can be seen working through to a final pure (if unattractive) form, just as Le Corbusier's designs for the church at Ronchamp, from his notebooks, evolve towards a modular and Modulor solution. For Loewy's designs, however, such comparative material is rare. Jeffrey Meikle points out that much of his 1930s and 1940s archive was destroyed in the 1950s, and the practice in his office of only keeping the final, approved sketch not only led, as has been remarked, to 'the best drawings ending in the waste paper basket' in Jay Doblin's words but also to the comparative material that would test the theory also disappearing. In some cases where there are multiple drawings, it is not always clear how to date them, or whether they are preliminary sketches or renderings made after final designs, and whether they are by Loewy himself – who admitted that he had no great personal talent for drawing – or by one of his colleagues. This is especially true of the NASA work, for example.

There is one series of drawings, however, which do show design in evolution. These are the sketches Loewy made to help the design team in their desert shack developing the Studebaker Avanti. They are accompanied by a series of snapshots of the quarter size clay model that was being worked on at the same time. (Even here, two of the supposed Avanti drawings reproduced in *Industrial Design* were in fact published in 1951, a decade before the project was on the drawing board.) Though it is difficult to place the drawings in a set order, it is easy to see how the design strove to get away from a traditional Detroit design – large radiator grilles, flared edges to panels, chrome highlights – towards a smoother, more European look. Certain features can be seen evolving, for example the flush sides and the inboard mounted headlamps, and the deliberate lift to the tail of the car. The gradual disappearance of the radiator grille can also be traced through the drawings and models. How, though, does this relate to the idea of MAYA? Clearly the design is pushing towards the client's brief for a 'different' two-seater sports car: the Avanti did break new ground in its styling, even if its handling, though stiffer than an ordinary American production car, was a long way from European sports car standards. Loewy offers one definition: 'there seems to be for each individual product a critical area at which the consumer's desire for novelty reaches what I might call the shock-zone.' Defining this shock-zone establishes the MAYA coefficient, but since the zone itself is subjective and arbitrary, all this definition does is establish that it is consumer acceptability that is most in view.

'Most advanced yet acceptable' does make more sense as a term applied to designs such as the S-1 Locomotive, where a dream of speed had to be harnessed to mechanical reality and to the railroad's aims to attract business in a competitive market. More seriously, the MAYA idea could be applied to the new

Loewy often used these mocking photos of poodle and tailfins to emphasize the excesses of Detroit *(above)*: his own design brief to the Avanti designers included these drawings *(facing)* to help develop design concepts rapidly

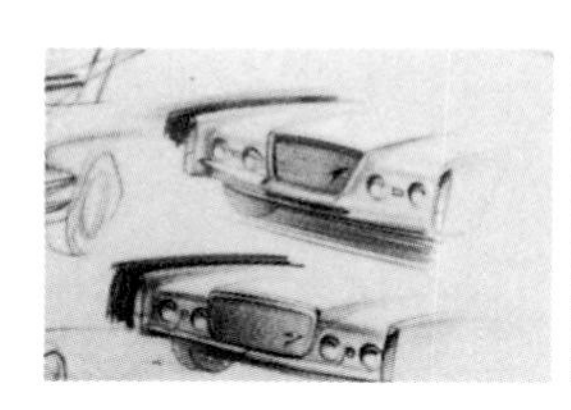

Work in the 1970s by Loewy's European offices included the TV camera for Thomson *(above)* by the Paris office, and the redesign of the Moskvitch car *(facing)* by the London office.

layouts for mechanised department stores such as Foley's, where research had shown a need to redesign the structure of the department store without upsetting the customer's perceptions of shopping. However, in considering Loewy's career and personality as a whole, the best application for the motto seems to be to the man himself. Loewy's insistence on the importance and relevance of industrial design somehow justifies his relentless self-advancement, if not his slightly opaque attitude to clear facts. Susan Heller, writing on Loewy's first fifty years as a designer in the *New York Times* Sunday magazine in 1979, calls him the pioneer of streamlining, just as the same paper a few years before spoke of the 'best-known industrial designer of his time'. Loewy had been in the advance guard, had set out the ground rules for the profession, had made the dazzling deals and had won the accolades. He intended to go on being there as long as he could, as well. At the age of 70 he took a high-speed driving

course with the American racing driver Caroll Shelby: seven years later he is roaring over Californian beaches in a dune buggy! The last years of his life sound almost like a fight for youth and longevity. He had always been up in front, and most advanced he was going to stay.

Where did all this relentless energy and drive come from? He had, by his own account, admittedly, lost his first fortune in the Wall Street crash, but two decades later he was a very successful man indeed, claiming in the 1949 *Time* article a salary of 200,000 dollars a year, when most American consumer products at the top of the range were intended for families generating 10,000 dollars a year. He knew the Kennedy family well – Jackie Kennedy asked him to design the Kennedy memorial stamp after her husband's assassination: the 'most carefully planned stamp in postal history' according to the US Postmaster General at the time of its issue. He had been a design ambassador to Japan in the 1950s and to Soviet Russia in the

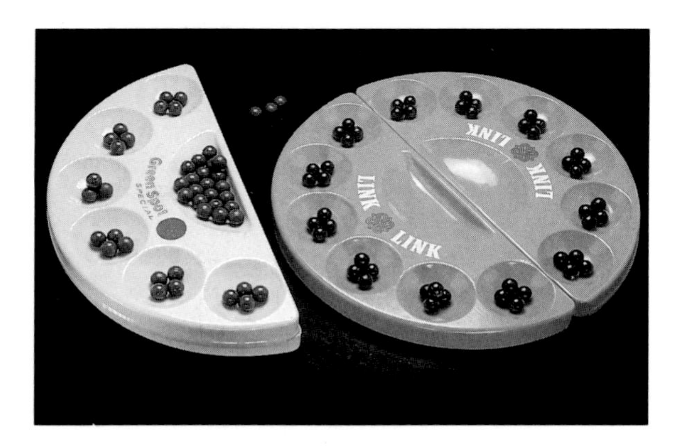

1970s – in itself a major first. His products were known around the world. His reputation, one would have thought, was assured, and retirement would have been well earned. But he wanted to carry on, making more and more claims, some of them quite outrageously wrong

The London office developed this oil carrier for Shell *(right)*, as well as other corporate identity and packaging design, while work for Philip Morris included a hospitality vehicle *(facing, above)* and promotional games *(below)*.

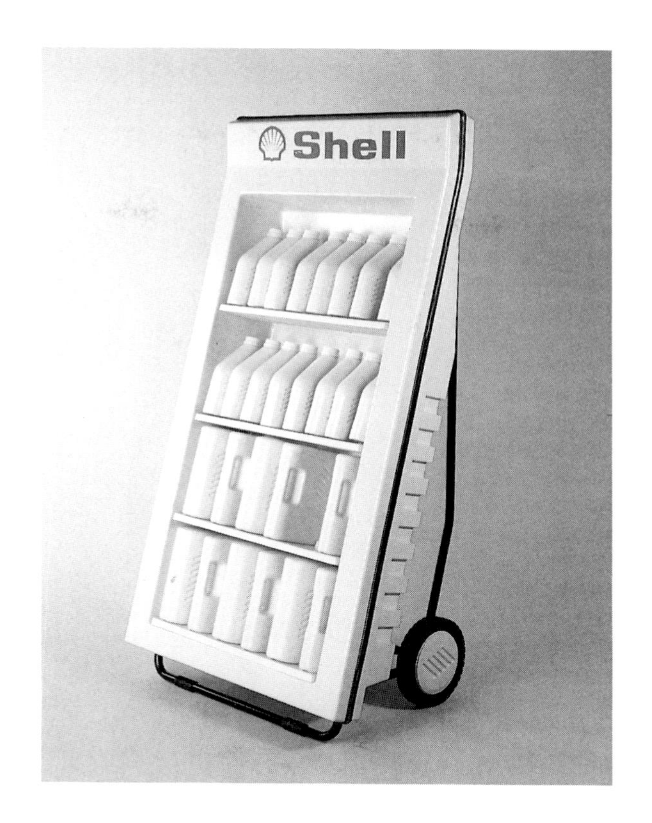

(for example over the design of the Coca-Cola bottle), and striving to keep himself always on the leading edge. His 'genial knack of putting himself to the front' seems to have infuriated many contemporaries and later commentators, content to write him off as a showman and nothing more. But his energy, even when misplaced, must command respect. One source for it that may have been overlooked is the fact of his surviving the First World War, despite having served all through it on the Western front. An English survivor of the First World War once told me that he felt at the end of the war that he could, and should, live for ever, and the feeling had never left him. Frank Williams' First World War memoirs are entitled *Old Soldiers Never Die*, not entirely without irony. Paul Fussell has written acutely of the sense of challenge and adventure the opening of post-war horizons brought to writers and artists in the 1920s. In Loewy's own case, the new horizon was the Manhattan skyline seen from the deck of his transatlantic liner. In *Never Leave Well Enough Alone* Loewy describes being taken one evening by the general manager of the Frigidaire plant to see the night shift arriving at the factory, 'Ain't it purty?' says his host: Loewy was 'utterly moved... It was like seeing the actual flow of the rich red blood of young, vibrant America.' It was a joie de survivre that in part helped keep Loewy most advanced.

Reviewing Loewy's 1951 book, Peter Blake, associate editor of *Architectural Forum* (and previously assistant curator at the Museum of Modern Art) noted Loewy's achievements, but sees him as a worried man 'due to a fundamental insecurity of taste'. The remark recalls Snaith's comment on Loewy's 'unerring vulgar taste'. The carping nature of the review – the book is a '100,000 word after dinner speech' – is not surprising: *Architectural Forum*, too, were cheerleaders for good design. Loewy, Blake wrote, 'lives in the special aura of a grade B movie and the Box Office is doing okay.' The double insult in the phrase is a telling one: not only was Loewy himself a phony, but he was a successful one – even less forgivable from a purist's viewpoint. Loewy, one is expected to realize, was beyond redemption: thus his insecurity of taste was indeed fundamental. To the 'good design' group, Loewy would never be acceptable. Philip Johnson said in 1979: 'Raymond started industrial design and the streamlined movement', thus jovially sidelining four later decades of work with a patronizing use of the Christian name: even his closest colleagues called him Mr Loewy.

His own contemporaries were not always too happy with him either. At a meeting of the Society of Industrial Designers in the mid-1950s, Loewy's name came up in the context of self-publicity. Henry Dreyfuss, who was present, commented 'We all know about Raymond Loewy (and publicity) but let's not forget that he is the best advertisement this profession has: without him the profession would be very different.' This may just be the traditional design profession closing ranks around an elder member, but the *Time* article in 1949 was certainly a boost for the profession, arguing as it did the case for the importance of industrial design as a management tool, and citing the work of other designers. Loewy saw his own pioneering work as good for the whole profession: writing shortly after the war, he said 'Today no manufacturer, from General Motors to the Little Lulu Novelty Company would think of putting a product on the market without benefit of a designer... This is in the space of twenty years. I take it as a complete vindication of my early theory that, eventually, correct visual presentation would become an integral part of merchandising in practically every field.' Yet his contemporaries did not seem to have warmed to him for taking the credit. As has been seen, Teague and Dreyfuss both wrote books that elaborated a considerable design philosophy – even if it was not the straight Bauhaus line favoured at MoMA. They were moving into place behind the need to have a theory. But the good design revolution, in short, passed Loewy by. The revolution claimed that design was much more than visual presentation as an integral part of merchandising: rather, it was design itself that gave

the product integrity. Design had moved from styling to become a core activity. It was not enough for the product to look right, the designer had to think right as well. Purity of form came from purity of thought, in good design terms.

In sum, the good design school imposed his 'uncertainty of taste' on Loewy. Looking back at it now, there was no uncertainty about it. His taste was, as Snaith said, unerring. It homed in on the expectations and unstated ideas of the average American, and when it hit home it was absolutely right. (When it missed it could be quite appalling.) What sustained Loewy, even under the attacks of the good design gang, was his knowledge that when he got it right it was right: the S-1 or the diesel was the way a streamlined locomotive should look. The Commodore Vanderbilt might look good too, but S-1 looked right. If Studebaker had followed the logic of the Starliner design, and put a robust engine into it, they would have en-

Loewy with the staff of the Paris office, the Compagnie de l'Ethésthetique Industrielle *(above)*: one of their main projects was the interior of the Air France Concorde *(facing)*.

joyed the success of the Mustang before Ford even twitched to the idea.

As Ada Louise Huxtable has pointed out, in the 1930s all sides believed in the perfectibility of society through design. The war years changed that perception, and the energy of the good design school increased. Without going as far as Tom Wolfe, whose *From Bauhaus to Our House* portrays the American architectural scene as enslaved to European models and elites, it is possible to perceive valid alternatives to the 'good design' approach which are more than justifications of styling. In the first place, the assumption that the logic of form following function as a necessity can now be seen as semantically flawed. There are plenty of pre-judgments in the concepts of form as meant by the good design guys. Form does not have to be symmetrical, classical or even itself logical. Natural forms, to take the immediate example, often break this rule. Nor does respect for manufacturing technology require particular forms: indeed one reason for the curves and curlicues of injection moulded products is simply that injection moulding works better with curved forms rather than squared ones. Above all, the didactic function of design, involving the insistence on principle and the denial of play, surprise or wit, is not a necessity. Good design pandered to an elite – not for all the rosewood and leather of such icons as the Eames chair – just as styling pandered to a wider denominator. The vernacular tradition of American design does not require European role models, just as there is a vernacular tradition of American architecture. Both traditions reflect the diversity of origins of American society, it is true. Whether this design tradition starts with the Franklin stove, the Colt Peacemaker or the Model T Ford, it has as its hallmark innovation and industrialization as its method. The geographical and demographical constraints in America involved the development of new forms of manufacturing and new forms of marketing. The existence of a large population, superficially united by the same language, political and economic system, produced opportunities for a mass culture unknown in Europe. This co-incided with an expansion in energy sources and technologies, and created a range of new products, for which a formal vocabulary had to be found. In some cases there were earlier European models: for the motor car and the revolver, for example, but American needs led to specifically American forms. It is in this tradition that Loewy holds a firm and important place. His contribution was at its most valuable during the 1930s and 1940s, and some of his petulance and striving for effect in later decades may reflect his own awareness that the torch was passing to younger hands. This does not excuse his errors – which sometimes seem willful omissions or mis-statements – but it may explain how they came about. Like many creative people, Loewy may not have been able to explain how his creativity worked – though he could talk at length about what he had done and would do. His wife Viola points out that he prepared presentations and negotiations meticulously, going into training in advance of important meetings with a special diet and extra exercise. He hated being caught off-guard, and so his act was always well prepared. In truth his ability as a showman, his endless taste – some would say greed – for publicity masked his creative streak.

In 1979 he complained glumly that young American designers were 'often influenced by fleeting styles. They're trying to be too damned intellectual. They're trying to make a pure science of industrial design.' Raymond Loewy died in France in 1987. Throughout his life as an industrial designer he had believed passionately in design and the importance of design. It formed all of his conversation, and everything turned around his work. When some of his drawings were put up for sale after his death, a group of American designers, led by the editor of *International Design* magazine, started a fund to buy some for the Library of Congress. Among the drawings is one made when Loewy was still a teenager in France, of a train, dated 1911. The bold signature – R Loewy – was still to be found on drawings sixty-five years later: by then he had created his own trains, cars, and products, giving a definitive shape to the American dream.

R
L

never slowed down since then

There is something about a Martini,

Ere the dining and dancing begin...

Ogden Nash

The story goes that at about midday one Friday the man in charge of the drinking-water coolers in the Loewy office removed the water bottles, empty or not, and replaced them with gallon containers of Martini, which was then available, to Loewy's preferred mixture and chilled as necessary, on tap, all afternoon. (The story as told to me did not, alas, resolve the question of whether Raymond Loewy mixed his gin and vermouth in six to one, ten to one or fourteen to one proportions.) But it must be said that there are reasons for doubting the story, if only on practical grounds, as anyone who has tried to remove a partly-full bottle from a water cooler will know. The story does not agree either with accounts of Loewy's serious approach to his office and his work. Indeed it is only as a caricature of Loewy that the story is interesting.

American management culture in the 1950s, as satirized in such films as Billy Wilder's *The Apartment*, did seem to endorse the Martini-based lunch. In turning the water cooler, a commonplace fixture, into a Martini fountain, Loewy (if the tale be true) encapsulated part of the American executive myth, while at the same time subverting it: the water cooler was the parish pump of office gossip, while the four-Martini lunch was a symbol of the officer class. But much as Loewy dearly loved a party, the

A caricature portrait of Loewy, composed of his signature.

idea of ending the week on such a note would not have appealed to him.

So the story as it stands must stay in doubt, because it is out of character, and also incomplete. Loewy certainly had the imagination and the verve to create splendid parties – for example, he tells in *Never Leave Well Enough Alone* of hiring, to perform at an office party, a young singer from Hoboken, someone called Frank Sinatra. The incompleteness of the story lies in the lack of any arrangement for providing the chilled glasses and the olive – or twist of lemon peel – without which a real Martini is an unfinished work. Loewy's unerring taste could have produced such a genial idea as the water cooler flowing with Martini, but if he were to carry it out he would have got all the details right – including getting the machine back to sobriety for Monday morning.

This story of the missing olive is perhaps a parable of Loewy's achievement. Certainly the person who told me the story mixed his admiration of the gesture with doubts about its acceptability – its propriety, almost. Charles Eames, the narrator seemed to be thinking, would never had done such a thing. The contrast between Eames and Loewy is not just one of generation, pre-war and post-war, and it goes deeper than the differences between the Eames's laid-back, craft-based approach to architecture and furniture and Loewy's joyous and pell-mell attack on product design. For many American designers and design writers the Eameses have a pure and noble quality, that is more than an appreciation of the high standards they set themselves and achieved in their work. It becomes a statement about what design ought to be, it idealizes design and the designer. One American designer – a robust and unsubtle believer in the work ethic – told me that simply seeing Ray Eames at a design

Loewy's first classic design, the Coldspot refrigerator.

meeting some years before her death brought tears of joy and admiration to his eyes. Loewy never inspired the same adulation among his contemporaries, and clearly this is not just a

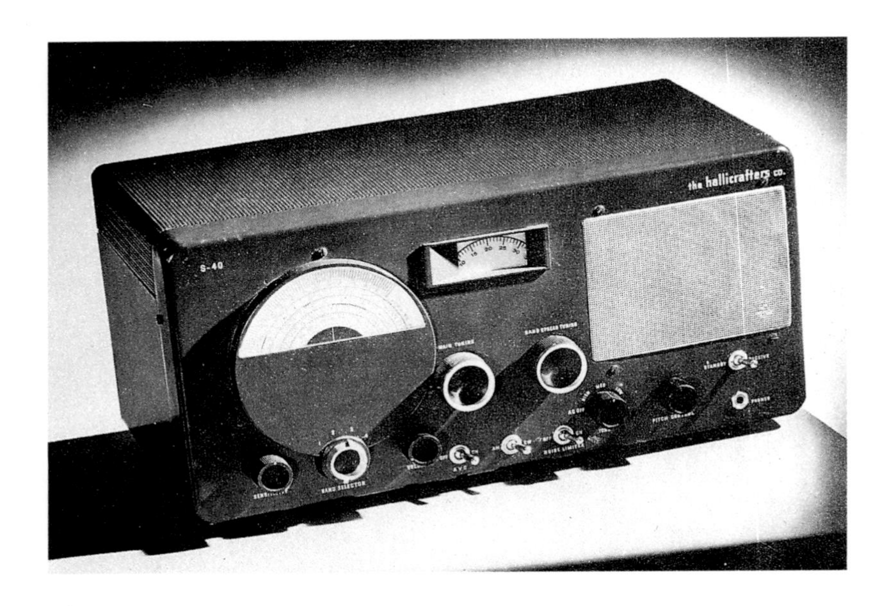

matter of charm and personality. Loewy had more than enough personality, after all, and those that worked for him speak of him with admiration and respect: his achievements have also been recognized posthumously by the American Society of Industrial Designers.

But the willful perception of Charles Eames as a father figure in American design effectively obscures his predecessors, and the main loser in this process has been Raymond Loewy. Although Loewy survived longer than his contemporaries (Charles Eames in fact predeceased Loewy, dying in 1978, while Dreyfuss died in 1972, Teague in 1960 and Bel Geddes in 1958) and so was available as a source when American design history was being written in the 1970s, his own enthusiasm for his own achievements created scepticism. It also may have seemed difficult to believe that one person had done so much, and had crammed so much into one lifetime – admittedly a long one – and was still at work.

But from the standpoint of today, Loewy's practical contribution to the development of American design can be assessed as important in two ways. Firstly there is his achievement as a product designer. From the mid-1950s until the 1970s a study group at the Illinois Institute of Technology, under the guidance of Jay Doblin, regularly asked designers which products they considered to be design classics. The first results – and all the subsequent ones – contained a number of Loewy products. In the selection finally published in the 1970s he is the American designer most widely represented (Charles Eames's furniture is next). This peer-group endorsement of Loewy's work includes the 1937 Coldspot Refrigerator, the Hallicrafter – 'the radio man's radio' – and the 1947 and 1953 Studebakers – the latter 'probably Loewy's greatest design' – as well as the Greyhound Scenicruiser bus. The overall list of classic designs is, by and large, a family of objects in common use, not a nostalgic reflection on pre-war glories or design esoterica. The actual choice emphasizes Loewy's claim to be

Two notable designs: the Hallicrafter radio *(facing)* and the 1947 Studebaker *(below)*.

considered among the prime fabricators of the American dream.

The second way in which Loewy must be considered important is in his promotion – by example and in his writing – of an independent industrial design profession. He shares with his immediate contemporaries the glory of having put product design to work in dragging American industry out of the Depression, and in developing the articulation of plenty that hallmarks post-war America, and in so doing he established the industrial designer as a necessary element in the development and marketing of sophisticated manufactured goods. But beyond that he was adamantly insistent about the independence of the designer. This is in part a reflection of his own restless versatility, but also an insight he had into the importance of design: like war, which cannot be left to the care of soldiers, design was too important to be left to the care of company men. Ironically, the main beneficiaries of his insight into the importance of design were to be the generation of Charles Eames (initially turned down for membership of the ASID because he worked in too narrow a field) and others who worked closely with individual companies, rather than ranging across the field of design. They owed much of their freedom of manoeuvre to the groundrules created by Loewy and his contemporaries.

After the Second World War Loewy wrote about the vindication of his wonderful dream of a world underpinned by design: the irony that haunted his last years was that the dream had come truer for others than for him. For all his successes and achievements in the post-war period, the newness was gone, and his best work was behind him. If he was aware of this, it did not break his spirit. Raymond Loewy remained a great enthusiast for design, and his books and his designs remind us that 'the de-

Riding through the American dream: Loewy in cowboy costume beside the first Starliner – with customary precision he noted that the hubcaps were wrong.

signer's life is an agreeable one: I do what I like doing'. He wrote in *Architectural Digest* in 1980 of coming to America some sixty years before: 'I had dreamed of ... calm and relaxation – and I found the Lexington Avenue Subway. But I quickly immersed myself in the American vortex, and I have never slowed down since then.'

author's note

Arthur J. Pulos's two books on American design (*see reading list*) are a touchstone and a model for others writing on the subject, and I am glad to have the opportunity of acknowledging here how much I have enjoyed them and learned from them. For the pre-war period, Donald J. Bush's book on streamlining is an equally important source. Loewy's own books provide both a stimulus and a challenge: our experience of Loewy would be much poorer without them. On Loewy himself, the recent exhibition catalogue, *Raymond Loewy, Pioneer of American Industrial Design*, edited by Angela Schoenberger, has been a guiding light, and I am grateful to the publishers, Prestel Verlag, for their permission to quote from it. Other sources are to be found in the reading list below, and in addition to thanking the authors and publishers concerned, I would also like to acknowledge the help and insights (the errors are my own) of Stephen Bayley, Robert Blaich, Jan Burney, Clive de Carle, Patrick Farrell, John Heskett, Roger Kennedy, Randolph McAusland, Ian Paten and Jurgen Tesch, among others, as well as thanking the library staff at the Design Museum, London.

Martin Pawley, the series editor, kept a firm hand on the tiller of this book's progress, and I am most grateful to him. Finally, my thanks are due to the Baronne de la Vourdiat for her hospitality while this book was being written.

Paul Jodard
October 1991

The author and the publisher wish to thank Raymond Loewy International for their help with the illustrations.

reading list

Albrecht, Donald *Designing Dreams* Harper & Row, New York 1986
Barr, Alfred H. (ed.) *Masters of Modern Art* MoMA, New York, 1954
Bayley, Stephen *The Conran Directory of Design* Villard, New York 1985
Bayley, Stephen *Harley Earl* Trefoil, London 1990
Bayley, Stephen *Sex, Drink andFast Cars* Faber, London 1988
Bel Geddes, Norman *Horizons* Dover, New York, 1977
Bel Geddes, Norman *Magic Motorways* New York, 1940
Bush, Donald J. *The Streamlined Decade*, Braziller, New York 1975
Caplan, Ralph *By Design* Harper & Row, New York 1982
Doblin, Jay *One Hundred Great Product Designs* Van Nostrand Reinhold, New York 1970
Dreyfuss, Henry *Designing for People* Paragraphic, New York 1967
Ferriss, Hugh *The Metropolis of Tomorrow*, New York 1929, Princeton Architectural Press, 1989
Forty, Adrian *Objects of Desire* Thames & Hudson, London 1986
Friedman, M. (et al) *Graphic Design in America*, exh. cat Minneapolis 1989
Heskett, John *Industrial Design* Thames & Hudson, London 1980
Holme, G *Industrial Design* The Studio, London & New York, 1934
Loewy, Raymond *Industrial Design*, Overlook, Woodstock, N.Y. 1979
Loewy, Raymond *The Locomotive*, London 1937, Trefoil 1987
Loewy, Raymond *Never Leave Well Enough Alone*, Simon & Schuster, New York 1951

Manchester, W. *The Glory and the Dream*, Michael Joseph, London 1973
Meikle, Jeffrey *Twentieth Century Limited* Philadelphia, 1979
Pawley, Martin *Buckminster Fuller*, Trefoil, London 1990
Pulos, Arthur J. *The American Design Adventure*, MIT Press Cambridge Mass., 1988
Pulos, Arthur J. *The American Design Ethic*, MIT Press, Cambridge Mass. 1983
Rand, Paul *Thoughts on Design* Van Nostrand Reinhold, New York (1949) 1970
Schoenberger, A. (ed.) *Raymond Loewy, Pioneer of Industrial Design*, exh. cat. (London, 1991), Prestel Verlag, Munich (contributions by Stephen Bayley, Donald J. Bush, Evert Endt, Reyer Kras, Jeffrey L. Meikle, Arthur J. Pulos, Elizabeth Reese, Bruno Sacco and others.)
Sexton, Richard *Classic Product Design* Chronicle, San Francisco, 1987
Teague, Walter D. *Design This Day* The Studio, London 1947
Terkel, Studs *The Good War*, Viking, New York 1984
Van Doren, Harold *Industrial Design* New York 1938
Wilson, Edward *Memoirs of Hecate County* Godine, Boston, 1980
Wolfe, Tom *From Bauhaus to Our House* New York, 1981
Wolfe, Tom *The Right Stuff* New York, 1979

Photographic credits by page:

Clive de Carle 83, 84, 87
Classic American 55, 95, 102
Design Museum 34, 35, 97, 99, 101, 119, 156, 161, 165
Evert Endt 62, 46,
General Motors 70, 164
Hulton Picture Co/Bettman Archive 31, 32, 68, 69, 105
Otto Kuhler 49
Library of Congress 14, 15, 32, 81, 158, 159, 183
Loewy International 2, 24, 25, 26, 29, 39, 50, 51, 79, 89, 93, 94, 99, 110, 121, 125, 141, 142, 144, 145, 146, 150, 155, 172, 173, 174, 175, 178, 179, 185
Vogue/ Condé Nast (USA) 64, 65, 66